Lecture Notes in Artificial Intell

Edited by J. G. Carbonell and J. Siekmann

Subseries of Lecture Notes in Computer Science

Bettina Berendt   Andreas Hotho
Dunja Mladenič   Giovanni Semeraro (Eds.)

From Web to Social Web: Discovering and Deploying User and Content Profiles

Workshop on Web Mining, WebMine 2006
Berlin, Germany, September 18, 2006
Revised Selected and Invited Papers

 Springer

Series Editors

Jaime G. Carbonell, Carnegie Mellon University, Pittsburgh, PA, USA
Jörg Siekmann, University of Saarland, Saarbrücken, Germany

Volume Editors

Bettina Berendt
Institute of Information Systems
Humboldt University Berlin, Germany
E-mail: berendt@wiwi.hu-berlin.de

Andreas Hotho
KDE Group at the University of Kassel, Germany
E-mail: hotho@cs.uni-kassel.de

Dunja Mladenič
J. Stefan Institute, Ljubljana, Slovenia
E-mail: Dunja.Mladenic@ijs.si

Giovanni Semeraro
Department of Informatics
University of Bari, Italy
E-mail: semeraro@di.uniba.it

Library of Congress Control Number: 2007934911

CR Subject Classification (1998): H.2.8, H.3-4

LNCS Sublibrary: SL 7 – Artificial Intelligence

ISSN   0302-9743
ISBN-10   3-540-74950-0 Springer Berlin Heidelberg New York
ISBN-13   978-3-540-74950-9 Springer Berlin Heidelberg New York

Springer is a part of Springer Science+Business Media

springer.com

© Springer-Verlag Berlin Heidelberg 2007
Printed in Germany

Typesetting: Camera-ready by author, data conversion by Scientific Publishing Services, Chennai, India
Printed on acid-free paper   SPIN: 12123469   06/3180   5 4 3 2 1 0

Preface

The World Wide Web is a rich source of information about human behavior. It contains large amount of data organized via interconnected Web pages, traces of information search, user feedback on items of interest, etc. In addition to large data volumes, one of the important characteristics of the Web is its dynamics, where content, structure and usage are changing over time. This shows up in the rise of related research areas like communities of practice, knowledge management, Web communities, and peer-to-peer. In particular the notion of collaborative work and thus the need of its systematic analysis become more and more important. For instance, to develop effective Web applications, it is essential to analyze patterns hidden in the usage of Web resources, their contents and their interconnections. Machine learning and data mining methods have been used extensively to find patterns in usage of the network by exploiting both contents and link structures.

We have investigated these topics in a series of workshops on Semantic Web Mining (2001, 2002) at the European Conference on Machine Learning / Principles and Practice of Knowledge Discovery from Databases (ECML/PKDD) conference series, in the selection of papers for the post-proceedings of the European Web Mining Forum 2003 Workshop, published as the Springer LNAI volume 3209 "Web Mining: From Web to Semantic Web" in 2004, as well as in the Knowledge Discovery and Ontologies workshop in 2004 and in the selection of papers for the post-proceedings of the ECML/PKDD 2005 joint workshops on Web Mining (European Web Mining Forum) and on Knowledge Discovery and Ontologies, published in 2006 as the Springer LNAI volume 4289 "Semantics, Web and Mining".

In 2006, we organized a workshop on Web mining that continues the aforementioned series of workshops on these topics. The workshop attracted a number of submissions and the highest-quality selected research papers, as well as the invited talk on "Web Usage Mining and Personalization in Noisy, Dynamic, and Ambiguous Environments" by Olfa Nasraoui (University of Louisville), fostered stimulating discussions among the participants. Specifically, the move from Web to Social Web (or Web 2.0) was an "emergent phenomenon" during the development of the workshop. The distinguishing mark of Social Web is user-generated content, which can play a key role if properly processed through advanced semantic technologies, such as text mining, natural language processing and image processing.

In fact, user-generated content represents a valuable source of information on users, in order to extract from content objects (bookmarks, blogs, photos, interaction logs, ...), relevant information about users (profiles) and the specific context in which they are interacting with a system, as well as to automatically annotate the content objects themselves and bootstrap the Semantic Web.

These topics were also investigated in the workshop "Ubiquitous Knowledge Discovery for Users" (UKDU) at ECML/PKDD 2006, which discussed the Web as one of today's most important ubiquitous environments. As the topics of that workshop complement the topics of our Web Mining workshop, this book also includes three invited and extended papers from the UKDU workshop.

Selected authors submitted expanded versions of their workshop papers. Those papers were reviewed again and the results of the selection were the eight papers chosen for this book.

The emergent phenomenon of Social Web and the widespread use of technologies such as Web logs, social bookmarking, wikis, RSS feeds are producing a significant change in Web usage. Understanding the dynamic of the relationship between topics and users in blogs, with the aim of constructing a plausible explanation for blogger behavior, is the main subject of the paper by Hayes, Avesani and Bojars. The paper proposes a set of measures to track topic and user drift, and shows how these measures can be used to explain user behavior. Collaborative environments are the basis of the Social Web. Flasch, Kaspari, Morik and Wurst consider the distributed organization of data employed in collaborative-filtering systems, which support users in searching and navigating media collections. They present Nemoz, a distributed media organizer based on tagging and distributed data mining.

The incorporation of semantics into the mining process is studied in two papers about Web usage mining. The invited contribution by Nasraoui and Saka provides a review of the recent efforts to incorporate content and other semantics to obtain a deeper representation of Web usage data, generally represented as a bag of clicks or URLs visited by a user. The paper examines the incorporation of simple cues from a Web site hierarchy in order to relate clickstream events that would otherwise seem unrelated. Facca concentrates on conceptual Web logs, that are XML documents enriched with information about the structure and content of the Web site. The paper shows how these logs can be automatically generated starting from a proper logging facility and a conceptual application model, and how this richer log representation allows one both to support the data mining process at different levels of abstraction and to analyze more easily the results of the mining process.

User profiles, as models of users' interests, play a key role in the recommendation of relevant content on the Web. Semeraro, Basile, de Gemmis and Lops describe a semantic recommender system able to provide the most interesting scientific papers to users according to their interests. The system learns semantic user profiles from documents represented using WordNet synsets. The hypothesis is that replacing words with synsets in the indexing phase helps learning algorithms to infer more accurate semantic user profiles. Anand and Mobasher, inspired by models of human theory developed in psychology, distinguish between users' short- and long-term interests; defining a recommendation process that exploits these two different models of users' interests. Often, the process of building user profiles relies on the analysis of digital data created or accessed by the users. The paper by Berendt and Kralisch focuses on other dimensions for understanding users' behavior: how language and culture may influence the way people access data and knowledge, and how these factors can be integrated into Web mining. A shift from technological to human aspects is needed for user-centered knowledge discovery, which deals with the ubiquity of people.

In the paper by Probst, Ghani, Krema, Fano and Liu, the authors propose an approach in which Web content (product descriptions) is processed in order to extract relevant attributes which can be used to describe items. The advantage of the approach is that it dynamically extracts attribute-value pairs, thus it differs from the classical information extraction task, in which a static template is filled in with relevant facts extracted from the text.

We thank our reviewers, the conference organizers, and the KDubiq project for sponsoring and support.

July 2007

Bettina Berendt
Andreas Hotho
Dunja Mladenic
Giovanni Semeraro

Organization

Web Mining (WebMine) 2006 was organized as part of the 17th European Conference on Machine Learning (ECML) and the 10th European Conference on Principles and Practice of Knowledge Discovery in Databases (PKDD).

Workshop Chairs

Bettina Berendt   Institute Information Systems
Humboldt University Berlin, Germany
Andreas Hotho   KDE Group at University of Kassel
Kassel, Germany
Dunja Mladenic   J. Stefan Institute
Ljubljana, Slovenia
Giovanni Semeraro   Department of Informatics
University of Bari, Italy

Program Committee

Sarabjot Anand	University of Warwick, UK
Mathias Bauer	DFKI, Germany
Janez Brank	J. Stefan Institute, Slovenia
Michelangelo Ceci	University of Bari, Italy
Marco de Gemmis	University of Bari, Italy
Miha Grcar	J. Stefan Institute, Slovenia
Marko Grobelnik	J. Stefan Institute, Slovenia
Pasquale Lops	University of Bari, Italy
Bamshad Mobasher	DePaul University, USA
Ion Muslea	Language Weaver, Inc., USA
Myra Spiliopoulou	Otto-von-Guericke-Univ. Magdeburg, Germany
Gerd Stumme	University of Kassel, Germany
Maarten van Someren	University of Amsterdam, Netherlands

Additional Reviewers

P. Basile   M. A. Torsello
I. Palmisano   D. Truemper

Table of Contents

An Analysis of Bloggers, Topics and Tags for a Blog Recommender System

Conor Hayes[1], Paolo Avesani[2], and Uldis Bojars[1]

[1] Digital Enterprise Research Institute,
National University of Ireland, Galway, Ireland
`conor.hayes@deri.org`, `uldis.bojars@deri.org`,
[2] ITC-IRST,
Via Sommarive 18
38050 Povo (Trento), Italy
`avesani@itc.it`

Abstract. Over the past few years the web has experienced an exponential growth in the use of weblogs or *blogs*, web sites containing journal-style entries presented in reverse chronological order. In this paper we provide an analysis of the type of recommendation strategy suitable for this domain. We introduce measures to characterise the blogosphere in terms of blogger and topic drift and we demonstrate how these measures can be used to construct a plausible explanation for blogger behaviour. We show that the blog domain is characterised by bloggers moving frequently from topic to topic and that blogger activity closely tracks events in the real world. We then demonstrate how tag cloud information *within* each cluster allows us to identify the most topic-relevant and consistent blogs in each cluster. We briefly describe how we plan to integrate this work within the SIOC[1] framework.

1 Introduction

A weblog (blog) is a website containing journal-style entries presented in reverse chronological order and generally written by a single user. Over the past few years, there has been an exponential growth in the number of blogs [14] due to the ease with which blog software enables users to publish to the web, free of technical or editorial constraints.

However, the decentralised and independent nature of blogging has meant that tools for organising and categorising the blog space are lacking. Advocates of the so-called Web 2.0 school of thought have proposed emergent organisational structures such as 'tag clouds' to tackle this problem. Tags are short informal descriptions, often one or two words long, used to describe blog entries (or any web resource). Tag clouds refer to aggregated tag information, in which a taxonomy or 'tagsonomy' emerges through repeated collective usage of the same tags.

---

[1] www.sioc-project.org — Semantically-Interlinked Online Communities.

B. Berendt et al. (Eds.): WebMine 2006, LNAI 4737, pp. 1–20, 2007.

In previous work we presented an empirical evaluation of the role for tags in providing organisational support for blogs [6]. In comparison to a simple clustering approach, tags performed poorly in partitioning the global document space. However, we discovered that, *within* the partitions produced by content clustering, tags were extremely useful for the detection of cluster topics that appear coherent but are in fact weak and meaningless.

We concluded that using a single global tag cloud as a primary means of partition is imprecise and has low recall. On the other hand, partitioning the blog document space using a conventional technique such as clustering produced multiple topic-related or *local* tag clouds, which could provide discriminating secondary information to further refine and confirm the knowledge produced by the clustering. Furthermore, local tag clouds established topic-based relationships between tags that were not observable when considering the global tag cloud alone.

This work was motivated by the need to build a blog recommender system in which a registered blogger would be regularly recommended posts or tags by other bloggers with similar interests. In such systems a key decision is how often the neighbourhood set or clustering needs to be calculated [12]. If similar users at time t are no longer similar at time $t+1$, models derived from data at time t may become obsolete very quickly.

We suggest a set of measures to track topic and user drift and we provide an explanation of topic evolution with reference to independently observed news events during the clustering period. Our initial results would suggest that many bloggers tend to have a short-lived attachment to a particular topic, which means that the neighbourhood relationships produced by each clustering cycle are relevant for a short period of time.

We then refine this analysis using information derived from the tag usage in each cluster. We find that blogs that contribute to the local tag definition of each cluster tend to be the most relevant in each cluster and, importantly, tend to be clustered together for extended periods. This behaviour suggests that topics uncovered by clustering have a core of relevant blogs surrounded by blogs that move between topics on a regular basis. In terms of defining a recommendation strategy, clustering followed by tag analysis allows us to define topics and potential authorities for those topics.

We briefly describe our current work which involves allowing the knowledge produced by automated learning techniques to be exported and reused using the SIOC (Semantically-Interlinked Online Communities) framework.

In the next section we give an overview of related work. Section 3 describes the datasets we use in this paper. Section 4 introduces our clustering method and the criteria we use for assessing cluster quality. In Section 4.2 we summarise our work on refining clusters using tag analysis. In Section 5 we introduce our experiments for tracking the relationship of users to topics as clustering is carried out on 6 data sets, each representing a week's worth of blog data. In Sections 5.1 and 5.2 we suggest a set of measures to track user and topic drift, and using these measures we provide an explanation of topic evolution in a cluster with reference to independently observed news events. In Section 6, we demonstrate how relevant sources of consistent topic-relevant information can be identified using simple tag analysis. We briefly describe our future work in Section 7 which involves integrating the information produced using knowledge discovery techniques with the SIOC framework. We present our conclusions in Section 8.

2   Related Work

The Semantic Web project has facilitated several initiatives concerned with linking and integrating topic-related material on the Web. For example, the SIOC framework facilitates the connection and interchange of information from Internet-based discussions and forums such as blogs, mailing lists, newsgroups and bulletin boards [1].

Tagging is a 'grassroots' solution to the problem of organising distributed web resources, with emphasis on ease of use. Quintarelli [10] proposes that tag usage engenders a *folksonomy*, an emergent user-generated classification. However, tags are flat propositional entities and there are no techniques for specifying 'meaning' or inferring or describing relationships between tags.

Although tagging is widely used by blog users, its effectiveness as a primary organising mechanism has not been demonstrated [2,6]. Despite its obvious weaknesses, tagging is firmly a part of the so-called Web 2.0 trend toward information sharing and collaboration on the Internet, typified by sites like the blog aggregator, Technorati[2], the photo-sharing site, Flickr[3], and the social bookmarks manager, Del.icio.us[4], all of which rely upon tags to allow users to discover resources tagged by other people.

Brooks and Montanez [2] have analysed the 350 most popular tags in Technorati in terms of document similarity and compared these to a selection of similar documents retrieved from Google. In previous work we have shown that the most popular tags form a small percentage of the overall tag space and that a retrieval system using tags needs to employ *at least* token-based partial matching to retrieve a larger proportion of tagged blogs [6]. Golder and Huberman [5] provide a good introduction to the dynamics of collaborative tagging on the Del.icio.us social bookmarks site. However, the Del.icio.us site differs from the blog domain in that tags are applied in a centralised way to URLs generally belonging to other people. A Del.icio.us user can view the bookmark tags already applied to the URL he wishes to index and choose an existing tag or use another. This aggregating facility is not available to the blogger, who must tag a piece of writing he/she has just completed. Whereas a tag on Del.icio.us references the URL of a website, a blogger's tag often references a locally defined *concept*.

Although the popular collective term 'blogosphere' implies a type of social network, recent research suggests that less-connected or unconnected blogs are in

---

[2] http://www.technorati.com
[3] http://www.flickr.com
[4] http://www.del.icio.us

the majority on the Web [7]. Link analyses on our datasets have produced the same results. For this reason we do not consider links between blogs in this paper.

3   Blog Data Sets

Our blog data set is based on data collected from 13,518 blogs during the 6-week period between midnight January 15 and midnight February 26, 2006[5]. All blogs were written in English and used tags. We found that blogging activity obeys a power law, with 88% of bloggers posting between 1 and 50 times during the period and 5% posting very frequently (from 100 to 2655 posts). On inspection, many of these prolific bloggers were either automated spammers ('sploggers') or community blogs. We selected data from 7209 bloggers who had posted from 6 to 48 times during the evaluation period. The median for this sample is 16 posts. On average, each user posted at least once per week during the 6-week period.

For each blog we selected the posts from the most frequently used tag during the 6-week period. This allowed us to associate a single topic (as defined by the blogger's tag) with each of the 7209 blogs. We chose to examine one topic per blog because blog topics from a single blog are often similar, as the blogger may use multiple tags for each post. Thus each of the 7209 blog 'documents' constitutes a single topic from a single blogger from the 6-week period.

The data was divided up into 6 data sets, each representing post data from a single week. As all 7209 bloggers do not post every week, the data sets have different sizes and overlap in terms of the blog instances they contain (see Table 1). Each instance in a data set is a 'bag of words' made up of the posts indexed under the most frequently used tag from a single blog during that week, *plus* the posts made in the previous 2 weeks (using the same tag). As the posts in a single week are often quite short and take the form of updates to previous posts, we include the previous 2 weeks to capture the context of the current week's updates. For example, if a blog is updated in week 3, the instance representing that blog in the dataset for week 3 is based on the posts in weeks 3, 2 & 1. If the blog is not updated in week 4, the instance representing the blog is excluded from the data set for week 4. As shown in Table 1, on average, 71% of the blogs present in the data set win_t will also be present in the data set win_{t+1}.

We processed each data set independently, removing stop words and stemming the remaining words in each document. We then removed low-frequency words appearing in less than 0.2% of the documents, and high-frequency words occuring in more than 15% of the documents. Documents with less than 15 tokens were not condsidered at this point. Each word was weighted according to the standard TF/IDF weighting scheme and the document vector normalised by the L^2 norm. This created a feature set of approximately 3,500 words for each data set. Table 1 gives the window period, size and overlap with the subsequent window.

---

[5] The blog URLs were kindly supplied by Natalie Glance of www.blogpulse.com

Table 1. The periods used for the windowed blog data set. Each period is from midnight to midnight exclusive. User overlap refers to the overlap with the same users in the data set for the next window.

data set	Dates (2006)	Size	# Feat.	Mean Feat.	Overlap win_{t+1}	%
win_0	Jan 16 to Jan 23	4163	3910	122	3121	75
win_1	Jan 23 to Jan 30	4427	4062	123	3234	73
win_2	Jan 30 to Feb 6	4463	4057	122	3190	71
win_3	Feb 6 to Feb 13	4451	4124	122	3156	71
win_4	Feb 13 to Feb 20	4283	4029	122.	2717	63
win_5	Feb 20 to Feb 27	3730	4090	121	-	-
mean	-	**4253**	**4043**	**122**	**3084**	**71**

4   Clustering and Tags

The blog domain contains many millions of documents, constantly being updated. A reasonable goal would be to try to organise these documents by topic or type. Document clustering is a well established technique for organising unlabelled document collections [15]. Clustering has two goals: to uncover latent structures that accurately reflect the topics present in a document collection and to provide a means of summarising and labelling these structures so that they can be interpreted easily by humans. Clustering has been used for improving precision/recall scores for document retrieval systems [11], browsing large document collections [3], organising search engine return sets [16] and grouping similar user profiles in recommender systems [13,9,8].

As our objective was to analyse user behaviour using a clustering solution, we implemented the *spherical k-means* algorithm, a well understood variation of the *k*-means clustering algorithm that scales well to large document collections and produces interpretable cluster summaries [4]. Spherical k-means produces *k* disjoint clusters, the centroid of each being a concept vector normalized to have unit Euclidean norm.

4.1   Clustering Quality

Given a set of data points, the goal of a clustering algorithm is to partition them into a set of clusters so that points in the same cluster are close together, while points in different clusters are far apart. Typically, the quality of a clustering solution is measured using criterion functions based on intra- and intercluster distance. Following [17], the quality of cluster r is given as the *ratio* of intra- to intercluster similarity, $\mathcal{H}_r$. Given S_r, the set of instances from cluster r, intracluster similarity, $\mathcal{I}_r$, is the average cosine distance between each instance, $d_i \in S_r$ and the cluster centroid, C_r. Intercluster similarity, $\mathcal{E}_r$, is the cosine distance of the cluster centroid to the centroid of the entire data set, C (see Equation 1).

$$\mathcal{H}_r = \frac{\mathcal{I}_r}{\mathcal{E}_r} = \frac{\frac{1}{|S_r|} \sum_{d_i \in S_r} \cos(d_i, C_r)}{\cos(C_r, C)} \tag{1}$$

In previous work, we have confirmed that clusters with high $\mathcal{H}_r$ scores tend to be clusters with large proportions of documents of a single class [6]

4.2   Partitioning by Tags or Clustering

A simple way to recommend new blog posts would be to use the tag label of each post to retrieve posts by other bloggers with the same tag. This is an approach used in a global tag cloud view of the blog domain. Tag clouds refer to aggregated tag information, in which a taxonomy or 'tagsonomy' emerges through repeated collective usage of the same tags.

Part A of Figure 1 illustrates this view of our blog data set. By clicking on a tag, the recent posts labelled with that tag are retrieved.

However, in any system where tags are aggregated, few tags are used very frequently and the majority of tags are used infrequently. This Zipfian tag-frequency

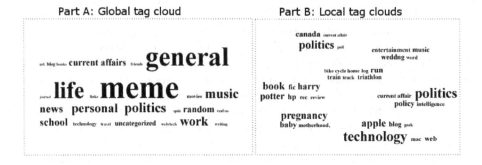

Fig. 1. Clustering produced multiple topic-specific tag clouds

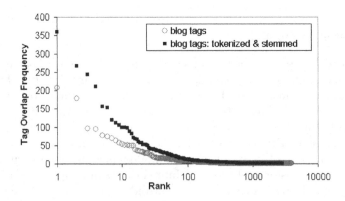

Fig. 2. Tag frequency vs. tag rank by frequency for the set of blog tags and blog tag tokens

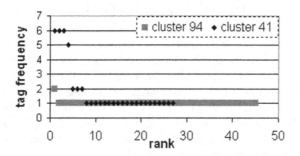

Fig. 3. Tag token frequency distribution for cluster 41 (high $\mathcal{H}_r$) and cluster 94 (low $\mathcal{H}_r$)

distribution means that only a small proportion of tags at any time can be used for retrieval purposes. Out of the 7209 documents in our data set only 563 (14%) out of 3934 tags were used 2 or more times, meaning that 86% of tags were useless for retrieval using an exact matching approach. This distribution is illustrated in Figure 2 where the circle icons represent 'raw' tag data and the square icons represent tags that have been tokenised and stemmed.

In previous work we demonstrated that tags generally performed poorly in comparison with clustering by content in identifying coherent topics in our blog corpus [6]. Furthermore, clustering by content partitioned the global tag space, producing multiple topic-related tag clouds as illustrated by Part B of Figure 1. In this view, the aggregated tag data in each cluster produced relationships between tags, which were not visible in the global view, and produced topic descriptions in the form of local tag clouds.

A key observation was that the tag frequency distribution per cluster varied according to cluster strength ($\mathcal{H}_r$). Weak clusters tended to have a long flat distribution, that is, few or no high-frequency tags (tokens) and a long tail of tags that have been used only once. Strong clusters tended to contain many high-frequency tags and a shorter tail.

Fig. 4. The tag clouds for cluster 41 (high $\mathcal{H}_r$) and cluster 94 (low $\mathcal{H}_r$)

Figure 3 illustrates the tag distribution for 2 clusters where k=100. Clusters 41 and 94 contain 47 and 43 instances per cluster respectively. Cluster 41 is in the top 20% of $\mathcal{H}_r$ scores and cluster 94 is in the bottom 20%. Figure 4 illustrates the tag cloud for each cluster based on these distributions. The tag cloud description of Harry Potter fan fiction shown in Figure 4 could not have been identified within the typical global tag cloud.

We refer to tag tokens that are not repeated by any other user in the cluster as **C-tags**. These tags are represented by the long tail of the frequency distribution and are not represented in the tag cloud view. **B-tags** are tag tokens with a frequency ≥ 2 that occur in several clusters at once. B-tags are analogous to stop-words, words that are so common that they are useless for indexing or retrieval purposes. Furthermore, b-tags also tend to be words with non-specific meaning, such as 'assorted', 'everything' and 'general'. As such, they do not contribute to cluster interpretation and are disregarded. **A-tags** are the remaining high-frequency tags. Clearly, a-tags are an important indicator of the semantics of the cluster as they represent an independent description of the cluster topic by 2 or more bloggers.

Combining clustering with subsequent tag analysis has allowed us to automatically identify and remove semantically weak clusters and to produce interpretable topic descriptions using local tag clouds [6].

5   Tracking User and Topic Drift

However, using clustering and tags on a static data set ignores the dynamic nature of the blogging domain. Blog data should be viewed as a stream of information, which we need to categorise and from which we need to extract the most relevant sources of information. The clustering solution we have described clusters blogs together by virtue of their similarity at a particular point in time. As bloggers continue to add new posts to their blogs, a key question is whether the relationships established by a clustering solution will be valid in the next time frame. Another key question is how the most relevant and consistent blogs associated with a particular topic can be identified.

In the following sections we attempt to make these questions clearer by measuring user and topic drift in our blog data over time. In the final section, we will turn again to tag analysis to allow us to identify bloggers that are consistently relevant to a given topic.

In these experiments we do not address the issue of selecting an optimal value of k and, as such, we cluster the data at several values of k. For each value of k, a random seed is chosen after which k-1 seeds are incrementally selected by choosing the seed with the greatest distance to the mean of the seeds already selected. In order to track user and topic drift from week to week, the seeds for the clusters in week t are based on the final centroids of the clusters produced in week t-1, except in the case of the first week, where the seeds are chosen to maximise interseed distance.

In order to cluster data using the seeds based on the centroids from the previous week we map the feature set from the previous week's data to the feature set

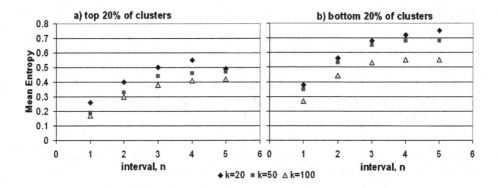

Fig. 5. Mean user entropy recorded where the intervals between windows vary from 1 to 5. The diagram on the left gives the entropy recorded for the top 20% of clusters according to $\mathcal{H}_r$. The diagram on the right gives the user entropy for the bottom 20% of clusters.

of the current week. In each pair of adjacent windows, the feature set overlap between windows is greater than 95%. The feature values for each seed are the feature weights from the corresponding centroid in the previous week.

In order to compare clustering in adjacent windows we define the following measures: *user entropy* per cluster, $\mathcal{U}_r$, and *interwindow similarity* per cluster, $\mathcal{W}_r$. User entropy, $\mathcal{U}_r$, for a cluster is a measure of the dispersion of the users in one cluster throughout the clusters of the next window. For a fixed value of k, if many of the users in a single cluster in win_t are also in a single cluster in win_{t+1}, then entropy will approach zero. Conversely, if the neighbourhood of users at win_t is spread equally among many clusters at win_{t+1}, entropy will tend toward a value of 1.

$$\mathcal{U}_r = -\frac{1}{\log q} \sum_{i=1}^{q} \frac{n_r^i}{n_r} \log \frac{n_r^i}{n_r} \qquad (2)$$

$c_{r,t}$ is cluster r at win_t; $c_{i,t+1}$ is cluster i at win_{t+1}, which contains users from $c_{r,t}$. S_{t+1} are all the instances in win_{t+1}. q is the number of $c_{i,t+1}$ (the number of clusters at win_{t+1} containing users from cluster $c_{r,t}$). $n_r = |c_{r,t} \cap S_{t+1}|$. n_r^i is $|c_{r,t} \cap c_{i,t+1}|$, the number of users from cluster $c_{r,t}$ contained in $c_{i,t+1}$.

The interwindow score, $\mathcal{W}_r^{t+1}$, for a cluster r in window win_t, is the similarity between the centroid of cluster r and the centroid of the corresponding cluster r in window win_{t+1}. Likewise, $\mathcal{W}_r^{t-1}$ is the similarity between the centroids of cluster r at windows win and win_{t-1}. Intuitively, $\mathcal{W}_r^{t+1}$ is a measure of the drift of the centroid concept, C_r, at win_t, where C_r is also the seed for cluster r at win_{t+1}.

$$\mathcal{W}_r^{t+1} = \cos(C_{r,t}, C_{r,t+1}) \qquad (3)$$

5.1   User Drift

In this section we examine whether users stay together as the data is clustered window by window. We demonstrate the degree of user drift by increasing the interval over which we calculate user entropy. We cluster the data in each window at $k =20$, 50 and 100, as described in the previous section. For each clustering we calculate user entropy for each cluster in win_t in relation to the clusters in window win_{t+n}, where the interval n is defined as $1 \leq n \leq 5$.

Table 2. The mean user entropies between adjacent window periods calculated over the top and bottom 20% of clusters ranked according to $\mathcal{H}_r$

$n=1$	$k = 20$				$k = 50$				$k = 100$			
	top 4		bottom 4		top 10		bottom 10		top 20		bottom 20	
Interval	$\hat{f}$	$\hat{u}$	$\check{f}$	$\check{u}$	$\hat{f}$	$\hat{u}$	$\check{f}$	$\check{u}$	$\hat{f}$	$\hat{u}$	$\check{f}$	$\check{u}$
0 - 1	0.1	0.34	0.35	0.48	0.16	0.24	0.29	0.42	0.19	0.2	0.27	0.28
1 - 2	0.1	0.28	0.39	0.37	0.13	0.21	0.35	0.38	0.15	0.17	0.32	0.27
2 - 3	0.1	0.26	0.41	0.37	0.12	0.2	0.37	0.37	0.13	0.19	0.39	0.3
3 - 4	0.11	0.24	0.40	0.37	0.12	0.15	0.40	0.3	0.1	0.15	0.38	0.22
4 - 5	0.11	0.19	0.43	0.32	0.08	0.15	0.39	0.27	0.11	0.14	0.44	0.25
Mean	0.1	0.26	0.4	0.38	0.12	0.19	0.36	0.35	0.14	0.17	0.36	0.26

However, rather than averaging the cluster entropy scores between windows over all clusters for a particular value of k, we examine 'strong' clusters (high $\mathcal{H}_r$) against weak clusters (low $\mathcal{H}_r$). Our hypothesis is that users associated with a strong cluster at window win_t will also tend to be together at window win_{t+1}. Conversely, we would expect greater user drift from clusters with low $\mathcal{H}_r$ scores. For the clustering produced at k in each window win_t we rank the clusters in descending order according to $\mathcal{H}_r$. For each pair of windows, win_t and win_{t+n}, we calculate $\hat{u}$, the average entropy of the top 20% of the ranked clusters in win_t, and $\check{u}$, the average entropy of the bottom 20% of the ranked clusters in win_t. We also calculate $\hat{f}$ and $\check{f}$, the respective fractions of the data set represented by the top and bottom 20% of the ranked clusters in win_t. Table 2 shows the $\hat{u}$ and $\check{u}$ scores for each pair of windows where $n = 1$. Figure 5 demonstrates the $\hat{u}$ and $\check{u}$ scores for $1 \leq n \leq 5$.

As a baseline, we observe that if the users in the clusters of win_t were randomly dispersed among the clusters in win_{t+n}, both $\hat{u}$ and $\check{u}$ would have values close to 1. On the other hand, if the users in each cluster of win_t were again clustered together in win_{t+n}, then both $\hat{u}$ and $\check{u}$ would have a value of 0. Table 2 shows that, even where n is low, the values for $\hat{u}$ indicate user dispersion from window to window. As the interval increases with n, $\hat{u}$ and $\check{u}$ also increase. This would suggest that the relationship between users, based on a shared topic, is short-lived rather long term. Comparing the values for $\hat{u}$ and $\check{u}$, it is clear that user drift is more pronounced for the clusters with low $\mathcal{H}_r$. However, Table 2 also shows that $\hat{f}$, the fraction of the data set contributing to the $\hat{u}$ score, is smaller

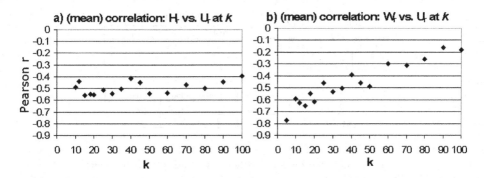

Fig. 6. a) Mean correlation between $\mathcal{H}_r$ and $\mathcal{U}_r$ as measured between pairs of windows at varying values of k. b) Mean correlation between $\mathcal{H}_r$ and $\mathcal{W}_r$ at k.

than $\check{f}$, the fraction of the data set contributing to $\check{u}$, by at least a factor of 2. This means that, in clusters where user drift is shown to be relatively low, the proportion of the total users involved is actually quite low.

By correlating $\mathcal{H}_r$ against $\mathcal{U}_r$ we can confirm the relationship between the two scores. For each pair of adjacent windows (win_t, win_{t+1}) we calculate the correlation of $\mathcal{H}_r$ against $\mathcal{U}_r$ at values of k from 5 to 100. We find a negative correlation for each of the 5 window pairs at every value of k. For each value of k we average the correlation scores produced from the 5 window pairs. Figure 6(a) graphs the mean correlation against k. The consistent negative correlation supports our observation that clusters with well defined topics (high $\mathcal{H}_r$ scores) are more likely to have less user drift (i.e. low $\mathcal{U}_r$) than clusters with low $\mathcal{H}_r$ scores. However, we should recall that clusters with high $\mathcal{H}_r$ consistently make up a small proportion of the overall data set.

5.2 Topic Drift

Topics do not remain stable over time. They emerge and decay or become transformed as lowly weighted features in one window are boosted in another window. Clearly, during this period of transition the relationship between users and clusters will be fluid. In order to demonstrate this we firstly examine the correlation between clusters in two adjacent windows in terms of their user entropy scores and their interwindow scores. Figure 6(b) demonstrates the mean correlation at k for $\mathcal{W}_r$ against $\mathcal{U}_r$ calculated between the clusters from the 5 pairs of adjacent windows. The strong negative correlation, particularly evident for $k < 50$, suggests that user drift is strongly related to concept drift.

5.3 Analysis

As bloggers add new posts they modify the topic description of the posts currently indexed under their tag. They also collectively modify the global topics that will be detected in the next clustering iteration. If users assigned to a cluster r in win_t post new material in win_{t+1} *dissimilar* to the cluster centroid of r,

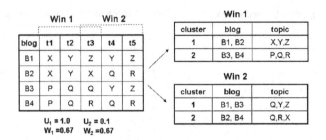

Fig. 7. A simplified example of how user and topic drift occur

then it is most likely that these users will not be associated with the same topic in win_{t+1}.

This type of behaviour is illustrated in the simple example in Figure 7 in which there are 4 blogs, {B1,B2,B3,B4}, each with five posts from the set P = {P,Q,R,X,Y,Z}. Each blogger has posted once at each time increment (t1 to t5). For the sake of simplicity we assume that the similarity between blogs is based on the proportion of overlap of elements from the set P. We cluster the blogs at time $t3$ and $t5$, using $k =2$. For each clustering, each blog is represented by the elements from P that fall within the period *win 1* and *win 2* respectively. In window *win 1*, the clusters produced are {B1,B2} and {B3,B4} and the respective cluster topic descriptions are {X,Y,Z} and {P,Q,R}. During window *win 2*, the bloggers B2 and B3 change topics, each selecting posts not associated with their cluster assignment from window 1, while bloggers B1 and B4 choose posts consistent with their cluster assignment. Clustering the data in window *win 2* produces the assignments shown in the bottom right of Figure 7. The clusters are {B1,B3} and {B2,B4}. This causes the user entropy $\mathcal{U}_r$ for each cluster in *win 1* to go to 1. We can also see that the topic descriptions in the clusters from *win 2* have been modified. The $\mathcal{W}_r$ score for each cluster is 0.67 where $\mathcal{W}_r$, in this case, is based on the proportion of overlapping elements. So in this example we can see topic drift between clusters is caused by bloggers moving away from the clusters they were assigned to in the first period. While this causes a large entropy score, we should also observe that the overall topic descriptions are changed but still have a degree of similarity with the topic descriptions produced in *win 1*. As such we suggest that although $\mathcal{U}_r$ and $\mathcal{W}_r$ are clearly related, the rate of topic drift may be considerably slower than the rate of user drift. We can see this from Table 3 where, after win_2, topic drift is extremely low ($\mathcal{W}_r \geq$ 0.93), while user drift is low but not negligible ($\mathcal{U}_r \geq 0.19$).

5.4   A Real World Example

In this section we provide an analysis of the relationship between user and topic drift based on independent empirical observation of news events.

In Table 3 we show the $\mathcal{W}_r$, $\mathcal{H}_r$ and high $\mathcal{U}_r$ scores for cluster 24 (from $k =50$) in each window. The cluster was chosen because it clearly illustrates a transition

Table 3. The change in W_r, H_r and U_r scores between windows as the 'Danish Newspaper Muhammad Cartoon Controversy' topic emerges. In each row the W_r and U_r scores refer to the drift since the *previous* window, win_{t-1}.

win_t	W_r^{t-1}	H_r	U_r	centroid key words	a-tags	nyt	wp	tg
win_0	-	0.41	-	sharon, bbc, mr, pilot, ariel	current, affair, politics, bsg, culture	0	0	0
win_1	0.78	0.44	0.39	sharon, israeli, palestinian, hamas, lee	current, affair, bsg, israel, politics	0	0	1
win_2	**0.28**	**0.87**	**0.66**	**muslim, cartoon, islam, danish, prophet**	politics, religion, current, affair, war	1	9	55
win_3	0.93	0.93	0.21	muslim, cartoon, islam, danish, prophet	politics, current, affair, war, society	13	14	42
win_4	0.96	0.96	0.19	cartoon, muslim, islam, danish, prophet	politics, current, affair, religion, culture	7	8	15
win_5	0.94	0.87	0.25	cartoon, muslim, islam, danish, prophet	politics, current, affair, religion, islam	7	5	4

from a weak to strong topic where the values W_r, H_r and U_r can be explained with reference to independent evidence.

In win_0 cluster 24 has a low H_r score and the most highly weighted terms from the cluster centroid suggest a cluster that may be mixing several topics. As the values of W_r and U_r refer to the difference between the previous window and the current window, we do not have these values for win_0. The topic descriptions in win_1 suggest that the topic has become more coherent, concentrating on Israeli/Palestinian affairs and the surprise win by the Hamas party in the Palestinian elections during win_1. The cluster has moderate similarity (0.78) to the previous week and a moderate level of entropy, suggesting that many users from the previous week have drifted away.

Win_2 brings a very large change. This is the week that the 'Danish Newspaper Muhammad Cartoon Controversy' began its month-long run in the world media[6]. By win_2, bloggers in our data set have begun to reference this issue and the topic immediately begins to dominate cluster 24. The cartoon controversy topic emerges from a weak cluster in win_1 ($H_r = 0.44$), describing events in the Middle East, to become a 'strong' topic ($H_r = 0.87$) in win_2. The rapid growth in H_r is accompanied by an equally rapid drop in W_r from 0.78 to 0.28, suggesting that the increase in H_r is due to the introduction of a stronger topic into the cluster. Furthermore, the U_r score at win_2 undergoes a large increase, suggesting that a large proportion of the users in cluster 24 at win_1 are no longer together in win_2. From win_2 to win_5, the cluster enters a stable period, with high H_r and W_r scores and lower U_r scores than before.

We can synchronise this behaviour with the real events. We suggested earlier that posts about the controversial election of Hamas in the Palestinian elections during win_1 had contributed to the increase in coherence of cluster 24. However,

---

[6] http://en.wikipedia.org/wiki/Jyllands-Posten_Muhammad_cartoons

with regard to the cartoon controversy we can be more precise. The columns marked *nyt*, *wp* and *tg* in Table 3 refer to the *New York Times*, *Washington Post* and *The Guardian* newspapers respectively. The numbers in the columns refer to the number of articles, commentaries and features carried by each newspaper about the controversy. To get these numbers we queried the archive sections of these newspapers using a query term extracted from the 5 most highly weighted terms in win_2 : 'muslim, cartoon, islam, danish, prophet'. From these figures, we can see that the emergence of this story in the international press is synchronised by its emergence in the blogosphere. Furthermore, we can construct a plausible explanation for user behaviour using the measures we defined.

5.5   A-Tag Meta-labels

Table 3 also illustrates that the a-tag descriptions of each cluster offer more abstract summaries of the cluster topic than those created by the cluster centroid. Furthermore, these tags often furnish information not apparent in the centroid descriptions. In this case, the term 'bsg' explains the poor $\mathcal{H}_r$ scores for the first 2 windows in Table 3. 'bsg' is an acronym for the cult science fiction TV show 'Battle Star Galactica', which has a central character called Sharon, a fighter pilot. Therefore, we can see that the centroid keywords for windows 0 and 1 refer to a set of documents concerning former Israeli prime minister Ariel Sharon, the Israeli-Palestinian conflict and a fictional account of war in space. The tag 'bsg' disappears in window 2 at the same time as user entropy increases dramatically, suggesting that the 'bsg' fans have moved from this cluster.

6   A-Blogs as Relevant Sources of Information

So far our analysis would suggest that the blog domain is characterised by bloggers moving frequently from topic to topic. Although strong clusters tend to have lower user entropy, these clusters form a small proportion of the overall data set. These observations would suggest that many bloggers tend to write in a 'shallow' way i.e. they are not regularly using terminology that allows them to be strongly associated with a particular topic for any length of time.

Furthermore, we have demonstrated the fluid relationship between bloggers and topics using a real world example of bloggers quickly reacting to an important breaking news story.

However, this analysis has not differentiated between blogs according to their relevance to the topic defined in the cluster. Intuitively, some blogs will provide information that is more relevant to the topic defined by the cluster than others. In the following section, we attempt to identify those blogs using tag analysis and we revisit the user drift experiments to see whether these blogs are more consistently associated with the same topics.

In Section 4.2 we described how each cluster can be described by a tag token cloud made up of a-tags. As the tag frequency distribution in each cluster follows a power law, only a portion of the blogs in each cluster will have contributed tag

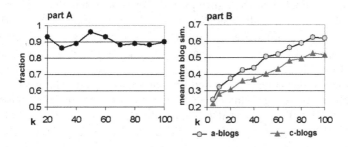

Fig. 8. Part A: mean fraction of clusters where a-blog IBS > c-blog IBS. Part B: mean IBS for a-blogs and c-blogs.

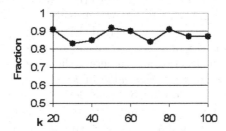

Fig. 9. The mean fraction of clusters where a-blog similarity to the cluster centroid > c-blog similarity to the cluster centroid. The mean is calculated based on the fractions obtained for each dataset at each value of k.

tokens to the tag description. For the sake of convenience, these blogs are termed *a-blogs*. The remaining blogs, which contribute single tag tokens to the long tail of the frequency distribution, are termed *c-blogs*. In this section we examine the characteristics of a- and c-blogs, keeping in mind our goal to automatically identify blogs that are most relevant to the topic definition produced by the cluster description.

The following experiments are based on clustering of each of the 6 blog datasets at values of k from 20 to 100. For each dataset and each value of k we chose the top 40% of clusters according to the clustering criterion $\mathcal{H}$. From this set, we removed any clusters identified as potentially weak or noisy by the cluster $\mathcal{T}_r$ score [6]. For each of the remaining clusters in each dataset, we measured the *intra-blog similarity* (IBS) of the a-blogs and the c-blogs. The IBS of a group of blogs is the mean pairwise similarity of all the blogs in the group, where similarity is measured using the cosine measure.

For the sake of space, the results presented in Figure 8 are averaged over the 6 datasets. Part A of Figure 8 gives the fraction of clusters at each value of k in which the IBS of the a-blogs was greater than the IBS of the c-blogs. Part B gives the mean IBS at each value of k. For each of the 6 datasets we found the difference between the means of the a-blog and c-blog scores to be significant at

0.05 alpha level. Part A of the figure provides evidence that in a high fraction of clusters a-blogs are generally 'tighter', that is, more similar to each other than c-blogs. Part B then illustrates the mean difference in IBS between a-blogs and c-blogs in each cluster at each value at k. From $k = 50$ upwards the difference is approximately 0.1.

In the second experiment we tested whether a-blogs were closer to the cluster centroid than c-blogs. The cluster centroid defines the 'concept' induced by the clustering process. The spherical k-means algorithm produces a weighted term vector where the weights reflect the normalised summation of the term weights contributed by the documents in the cluster. The documents in a cluster will have differing degrees of similarity to the cluster centroid. Document vectors close to the centroid are more likely to contain highly weighted terms that are also highly weighted in the centroid vector. As such we would expect documents close to the centroid to be highly relevant to the concept description. Using the same set of clusters from each dataset, we measure the mean similarity of the a-blogs and c-blogs to each cluster centroid. Figure 9 presents the fraction of clusters where the similarity of a-blogs to the cluster centroid is greater than the similarity of c-blogs. The fraction shown here is the mean based on the fractions obtained from each dataset at each value of k. The figure indicates that the a-blogs in each cluster are more likely to be closer to the cluster centroid than c-blogs. Figure 10 illustrates mean similarities to the cluster centroid for a-blogs and c-blogs for each of the 6 datasets. For each dataset the difference between the means of the a-blog and c-blog results was found to be significant for each dataset at an alpha level of 0.05.

The results from these first two experiments lead us to conclude that within each cluster a-blogs tend to form tight subgroups, which are generally more similar to the cluster centroid than the remaining c-blogs in the cluster. A key

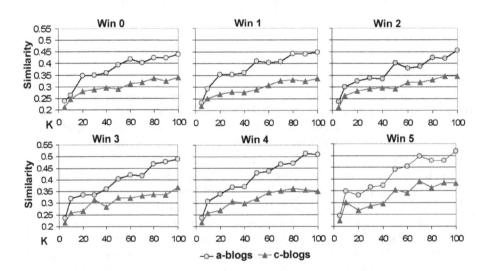

Fig. 10. The similarity to the cluster centroid for a-blogs and c-blogs

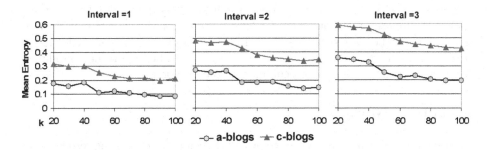

Fig. 11. Mean a-blog vs. c-blog entropy at interval $= 1,2 \& 3$

question is whether a-blog documents are more *relevant* to the cluster concept than c-blog documents. In information retrieval the cluster hypothesis [11] posits that documents that are more similar to each other are more likely to be relevant to a particular information requirement than less similar documents. The information requirement in this case is the concept summary presented by the cluster. In application terms, this is a synopsis of the topic presented to the user based on selection of key words and the retrieval goal is to suggest a set of blogs that are most relevant to the concept summary.

6.1   User Entropy Revisited

In previous work on the same datasets we described the phenomenon of *user drift*. This refers to the observation that, as the datasets are clustered from one week to the next, many blogs are often not clustered together again. This is problematic as it suggests that blog data requires constant re-clustering and that the relationships established between blogs based on shared topics in one week cannot be exploited for any length of time. It also suggests that (many) bloggers may be writing in a 'shallow' way i.e. they are not regularly using terminology that allows them to be strongly associated with a particular topic.

However, our previous analysis did not differentiate between blogs in each cluster and the entropy measure was calculated over both a-blogs and c-blogs. We return to this experiment and calculate the entropy for a-blogs and c-blogs separately in each cluster. Using the same clusters as before, the mean entropy is calculated at different values of k where the interval between datasets is increased from 1 to 3. For example, when the interval is 1 we calculate the mean entropy based on the entropy scores recorded between the following pairs of windows: (win_0, win_1), (win_1, win_2), (win_2, win_3), (win_3, win_4) and (win_4, win_5). When the interval is 3 the mean entropy score is based only on the following pairs : (win_0, win_3), (win_1, win_4) and (win_2, win_5). Figure 11 illustrates that a-blogs have much lower entropy than c-blogs at all values of k. As the distance between windows (and each clustering) increases, we would expect to see a rise in entropy. However, a-blogs have significantly smaller entropy scores and experience smaller increases in entropy than c-blogs as the interval increases.

This is an important observation because it suggests that not only do a-tags allow us to identify relevant sources of information about a topic, but that these sources tend to be *consistent* over time. In other words, we can identify bloggers that are consistently associated with topics and would be important candidates to consider in any topic-based recommendation strategy.

7   Future Work

The motivation behind this work is to provide a means of interlinking resources so that users can find and use relevant topic-related material from several sources. As such, we share many of the same goals of SIOC, an open-standard machine readable format for expressing and linking the information contained both explicitly and implicitly in Internet discussion methods such as blogs and bulletin boards [1].

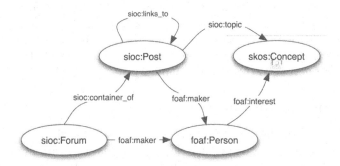

Fig. 12. Main concepts and connections in the SIOC data model

The basis for SIOC is an RDF-based schema which describes the main concepts found in online communities. The SIOC ontology is defined using the RDF/OWL language, which makes it possible for software to process the objects and relations described in a SIOC document. A SIOC document, unlike a traditional Web page, can be integrated with other Semantic Web documents to create a unified database of information. Figure 12 shows a high-level view of the SIOC data model. SIOC is used in conjunction with Dublin Core (DC) vocabulary for defining additional properties; FOAF vocabulary for describing information about people and their relations; and SKOS to describe categories and tags. A number of sites, such as LiveJournal, use RDF and FOAF to provide information about creators of the content, which can then be integrated with blog content data. Using properties such as the sioc:topic or the sioc:creator_of, content can be collected across many discussion platforms based on what is being talked about and who is saying it.

As such, our current work involves building services based on knowledge discovery that will enrich the machine-readable representation of blog data. An application of this is illustrated in Figure 13. One problem faced by bloggers is

Fig. 13. Simple SIOC-enabled tag recommendation service

the selection of meaningful tags after a post is written. A tag recommendation service accepts SIOC metadata about a blog post as input and returns a selection of tags the blogger might use. In this case the recommendation service is based on the clustering techniques we have demonstrated in this paper. The input post is matched to the most similar cluster and a selection of a-tags returned to the user. A secondary service would also return a list of a-blogs to the input post. By selecting one or several a-tags the blogger ensures that his/her post can be more easily matched or found by other bloggers or even readers from community sites such as bulletin boards. This is one of the goals of the SIOC framework. Using the SIOC export plugin the blogger can export his post metadata enriched with information received from the recommendation service.

8   Conclusion

In the analysis presented in this paper, each blogger was represented by a single 'topic', extracted from his/her most frequently used tag. We created 6 data sets, each representing a week's worth of data from the user's tag. We proposed a set of measurements for measuring user and topic drift and we demonstrated how they can be used to construct a plausible explanation for user behaviour. We found that the blog domain is characterised by many bloggers moving frequently from topic to topic. These observations would suggest that the majority of bloggers tend to write in a 'shallow' way about a variety of different subjects. We demonstrate the fluid relationship between bloggers and topics using a real world example of bloggers quickly reacting to an important breaking news story. We then demonstrated how we use tag information to refine the output of a clustering solution. We suggest that a-bloggers, bloggers who contribute tokens to the cluster a-tag description, tend to be the most relevant sources of topic information. Our evaluation found that a-bloggers tend to form the core of each cluster. Furthermore, we demonstrated that these bloggers tend to be clustered together again in later periods. Finally, we gave a brief introduction to our current work which involves integrating clustering and tag analysis with the SIOC framework.

Acknowledgments

This material is based upon work funded by project 8609: PeRec of the Provincia Autonoma di Trento and Grant No. SFI/02/CE1/I131 of Science Foundation Ireland.

References

1. Breslin, J.G., Harth, A., Bojars, U., Decker, S.: Towards semantically-interlinked online communities. In: 2nd European Semantic Web Conference, pp. 500–514 (May 2005)
2. Brooks, C.H., Montanez, N.: An analysis of the effectiveness of tagging in blogs. In: 2005 AAAI Spring Symposium on Computational Approaches to Analyzing Weblogs, pp. 9–14. AAAI, Stanford, California, USA (2005)
3. Cutting, D.R., Karger, D.R., Pedersen, J.O., Tukey, J.W.: Scatter/gather: a cluster-based approach to browsing large document collections. In: 15th international ACM SIGIR conference, pp. 318–329. ACM Press, New York (1992)
4. Dhillon, I., Fan, J., Guan, Y.: Efficient clustering of very large document collections. In: Grossman, R., Kamath, a.R.N.G. (eds.) Data Mining for Scientific and Engineering Applications, Kluwer Academic Publishers, Dordrecht (2001)
5. Golder, S.A., Huberman, B.A.: The structure of collaborative tagging systems. Journal of Information Science 32, 198–208 (2006)
6. Hayes, C., Avesani, P., Veeramachaneni, S.: An analysis of the use of tags in a blog recommender system. In: IJCAI 07, Hyderabad, India, pp. 2772–2777 (January 2007), http://www.ijcai-07.org
7. Herring, S., Kouper, I., Paolillo, J., Scheidt, L.: Conversations in the blogosphere: An analysis "from the bottom up". In: Proceedings of HICSS-38, p. 107. IEEE Computer Society Press, Los Alamitos (2005)
8. Kelleher, J., Bridge, D.: An accurate and scalable collaborative recommender. Artificial Intelligence Review 21(3 - 4), 193–213 (2004)
9. O'Connor, M., Herlocker, J.: Clustering items for collaborative filtering. In: ACM SIGIR Workshop on Recommender Systems, Berkeley, CA (1999)
10. Quintarelli, E.: Folksonomies: power to the people. paper presented at ISKO Italy-UniMIB Meeting, Mi (June 2005)
11. Rijsbergen, C.J.V.: Information Retrieval. Butterworth-Heinemann, Newton, MA, USA (1979)
12. Sarwar, B., Karypis, G., Konstan, J., Reidl, J.: Item-based collaborative filtering recommendation algorithms. In: 10th international conference on WWW, pp. 285–295. ACM Press, New York (2001)
13. Sarwar, B.M., Karypis, G., Konstan, J., Riedl, J.: Recommender systems for large-scale e-commerce: Scalable neighborhood formation using clustering. In: 5th International Conference on Computer and Information Technology (2002)
14. Sifry, D.: State of the blogosphere, april 2006 part 1: On blogosphere growth (2006) http://technorati.com/weblog/2006/04/96.html
15. Steinbach, M., Karypis, G., Kumar, V.: A comparison of document clustering techniques. In: 6th ACM SIGKDD World Text Mining Conference, Boston (2000)
16. Zamir, O., Etzioni, O.: Grouper: A dynamic clustering interface to web search results. In: 8th International WWW Conference, Toronto, Canada (May 1999)
17. Zhao, Y., Karypis, G.: Empirical and theoretical comparisons of selected criterion functions for document clustering. Machine Learning 55(3), 311–331 (2004)

Combining Web Usage Mining and XML Mining in a Real Case Study

Federico Michele Facca

Dipartimento di Elettronica e Informazione,
Politecnico di Milano, Italy
facca@elet.polimi.it

Abstract. In this paper we report our first extended experiments on Conceptual Web log generation and XML Mining over generated Conceptual logs. Conceptual logs are XML Web server log containing rich information about the structure of a Web site and its content. Furthermore they can be automatically generated starting from a proper logging facility and a conceptual application model. This allows an easier analysis of the results of the mining process, thanks to the rich information provided and allows to perform the data mining process at different levels of abstraction. In this work we use WebML as conceptual model, and XMINE as mining tool; nevertheless the underlying idea is of general validity and can be applied to any other conceptual modeling framework and mining technique.

Keywords: Web Usage Mining, XML Mining, XML Conceptual log, WebML, XMINE.

1 Introduction

Web Usage Mining, often referred as Web Log (Server) Mining, aims at extracting knowledge from Web servers logging facilities. Many research papers have been published on it and many commercial tools have recently reached maturity. At the same time XML is becoming widely used on the Web; nevertheless the research in the area of XML Mining is still at the first steps and few real case studies have been proposed and analyzed. In this paper we perform a data mining task over a rich XML Web log. Therefore, we adopt XML Mining techniques in the area of Web Usage Mining. In particular we collected more than 20,000 user sessions from the Web site of the Computer Science department of our university[1]; then after first experiments to test existing to for XML mining, we adopted XMINE [1] – a tool to mine rules from XML data – and performed the XML Mining task with it. Various experiments and researches have been conducted in this field in the last decade; most of these researches evidenced that the most demanding task is the analysis of the results produced by the data mining process. This is mostly caused by the poor information contained in the logs about the real content browsed by users.

---

[1] http://www.elet.polimi.it

B. Berendt et al. (Eds.): WebMine 2006, LNAI 4737, pp. 21–40, 2007.

Cooley [2] showed that, not only is the Web Usage Mining process enhanced by content and structure knowledge, but it cannot be completed without it. Hence data preprocessing becomes one of the fundamental task to improve the results of the Web Usage Mining task and to simplify their analysis. Only few researches on Web Usage Mining deal with the problem of enriching the information contained in the Web logs to improve the quality of the extracted knowledge. Other Web Mining techniques can be combined together with Web Usage Mining to actually solve the problem: i.e., Web Content Mining can be used to retrieve and model the content associated to the each navigated URL; Web Structure Mining can be applied to compute the structure of the Web site by following the links that interconnect its pages. This approach is time and computational demanding; it requires further expert analysis to validate results and off course a number of errors can be introduced in the reconstruction of contents and structure. Stumme et al. [3] address the problem using Semantic Web techniques to add knowledge about the page content to the Web log; in [4] pages navigated by users are tagged with keywords extracted from themselves; Punin et all [5] enrich Web log information adding Web site's maps. A complete survey on Web Usage Mining research can be found in [6].

Here we follow a different approach: we developed a framework to automatically generate enriched Web logs from the conceptual model of the application. This approach was first introduced in [7]. This work follows the way paved by [7] conducting the first extended experiments of enriched XML log generation and Web Usage Mining on generated logs. The approach is based on the Web Modeling Language (WebML) [8] and its supporting CASE tool WebRatio [9], for the design and the development of data-intensive Web applications. However, the illustrated results are of general validity and apply to any application that has been designed using a model-driven approach, provided that the conceptual schema is available and the application runtime architecture permits the collection of customized log data.

This paper is organized as follows: Section 2 summarizes the current efforts in the area of XML Data Mining and introduces the tool we used for the XML Mining task. In Section 3 we introduce the XML Conceptual logs. Section 4 presents the analyzed Web application and Section 5 reports some of the evidences we found mining the Conceptual logs. Finally, in Section 6, we address future research efforts and draw some conclusions.

2   XML Data Mining

The eXtensible Markup Language (XML) has rapidly become an important standard for representing and exchanging information through its applications. With the dramatic increase of information available in XML, there is a pressing need for languages and tools to manage collections of XML documents, as well as to mine interesting information from XML document collections.

The XML data mining research can be divided in two main areas: mining frequent pattern from XML data [1,10] and classifying XML data [11,12]. The literature shows that, despite the fact that XML is more and more used and a large number of XML documents is available, most of the researches deal with classifying XML data, disregarding that data classification, can efficiently performed only starting from data pattern. Our focus is on the few researches that somehow aim at extracting frequent patterns from XML data, as these are the most used techniques in Web Usage Mining. First studies in this area used techniques derived from *Text Mining*. Text Mining is an area of data mining that focused on finding repeating patterns inside text databases, i.e. Text Mining find frequent pattern of words inside a collection of phrases. In this framework an XML document is considered as a bag of words, and patterns are extracted from such bag [13].

A second approach, *native XML data mining* is quite new to the area of data mining. As far as we know, most of the studies in this area focus on mining frequent trees inside XML files. Tree mining over XML was first proposed in [10], a similar approach is also used in [11,14]. In [1] a language to extract association rules from XML documents, XMINE, is proposed and extended in to mine sequential patterns in [15]. We define approaches like the one introduced by Zaki [11], that finds all the frequent tree structures repeating in a collection, *brute force* approaches, while approaches like the one in [15], that finds only the frequent patterns corresponding to a certain XML structure, *structure-based* approaches.

As a working example, we introduce the XML document depicted in Figure 1 which represents various information about a department. In particular, it stores information about the available Ph.D. courses (identified by the tag `<PhDCourse>`) and about the people in the department (`<People>`). These can be either students (`<PhDStudent>`) or professors (`<FullProfessor>`). For each of them, some personal information are stored (`<PersonalInfo>`) as well as the list of works published (`<Publications>`) such as books (`<Book>`), journal papers, or conference papers (`<Article>`).

An XML document can be represented with a tree: each tag of a XML document can considered as node of the tree if it has other subelements or as a leaf if it has no other subelements (see Figure 2).

2.1 Tree Mining over XML Data

Zaki [11] proposes a method to find frequent structures within XML documents in order to classify them, i.e., a set of preclassified XML documents (*training dataset*) is used to develop a model to classify XML documents (*test dataset*) that still do not belong to a class. The model is created using the underlying structure of the preclassified documents. We focus our interest only on the frequent structure mining method presented in the article. The definition of the problem of mining frequent trees within an XML document is trivial as it is very similar to the original generic problem of mining frequent subtrees in a tree. In fact an XML document is a labeled rooted tree: i.e., a tree with a top element (the *root*) and where each node has a *label*. Let us consider the XML document in Figure 1: we're interested in finding XML fragments like `<Article><Author/></Article>`

```
<DEPARTMENT>
 <PhDCourses>
  <Course teacher="fp1" title="Advanced Data Mining">
   <TimeTable>...</TimeTable>
   <Student ref="ps1" />
   <Student ref="ps2" />
  </Course>
  <Course teacher="fp3" title="Intricacies of XML parsers">
   <TimeTable>...</TimeTable>
   <Student ref="ps2" />
   <Student ref="ps3" />
  </Course>
 </PhDCourses>
 <People>
  <PhDStudent id="ps2" advisor="fp3">
   <PersonalInfo email="fp3@cs.atlantis.edu">
    <Name>...</Name>
   </PersonalInfo>
   <Subscription year="2001" />
   <Publications> ... </Publications>
  </PhDStudent>
  <FullProfessor id="fp3">
   <PersonalInfo email="fp3@cs.atlantis.edu">
    <Name> ... </Name>
   </PersonalInfo>
   <Publications>
    <Article title="Golden Data Mines in Atlantis">
     <Author>Wilson</Author>
     <Author>Holmes</Author>
     <Conference name="VLDB" year="2001" />
    </Article>
    <Article title="P is just like NP - The Final Proof">
     <Author>...</Author>
     <Journal year="2000" month="4" volume="4"
              name="DMKD" publisher="Kluwer" />
    </Article>
    <Book year="2001" title="XML Query Languages">
     <Author>...</Author>
     <Publisher>...</Publisher>
     <Keyword>XML</Keyword>...<Keyword>XQuery</Keyword>
    </Book>
   </Publications>
   <Award year="2001" society="IEEE">This award..</Award>
  </FullProfessor>
 </People>
</DEPARTMENT>
```

Fig. 1. http://www.cs.atlantis.edu/research.xml, a sample document with various information about the research activities of a university department

or <Publications><Author>Holmes</Author></Publications>. We have not defined a particular task, we simply want to find all the possible subtrees in the starting document and among them, the most frequent subtrees. E.g. the fragment <Article><Author/></Article> means that there are a number of XML subtrees with a node <Article> that contains at least one node <Author>. Hence we are interested in finding embedded and unordered subtrees: i.e., subtrees derived not only considering the direct parent-child branches, but also ancestor-descendant branches, and where the order of children has no importance.

The algorithm applied by Zaki to find frequent substructure is the one proposed in [16]. To apply the algorithm, the XML document is represented by

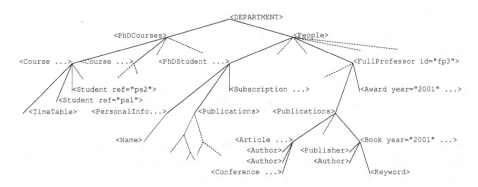

Fig. 2. A tree representation of the document in Figure 1

its *string encoding* denoted τ, obtained adding vertex numeric labels to τ in a depth-first preorder trasversal of $\mathcal{D}$ and adding a unique symbol $-1 \notin L_x$ whenever we backtrack from a child to its parent.

Tan et all [14], propose a method related to the one of Zaki. In particular, the problem of mining directly XML documents without replacing literal labels with numerics ones: the notion of string encoding is extended to *xstring encoding* to describe an XML document without loss of both structure and semantics.

The approach presented in this paragraph to mine frequent patterns in XML data, due to his nature is more suited to find tree patterns including only structural elements. In fact, compared to these ones, text nodes and attribute values, have a very small support; this is due to the fact that this approach aims to find all possible frequent patterns in a brute force fashion, and off course, structural elements are much more frequent than actual values of text nodes and attribute values.

2.2   Mining Rules from XML Data

Braga et al. [1] proposes a language to mine association rules from XML data. The language is later extend in [15] to sequential rules. In this work also a formal definition of the two problems is given.

Nevertheless the complexity of its formalism, thanks to the introduction of a language to express the mining task by means of XPath and XQuery, XMINE can be easily used by XML experts. The language proposed is based on the assumption that information like DTD and XSD schema can simplify the data mining task reducing the problem by providing proper constraints on the structure of the expected resulting pattern.

For example, let us consider an XML document in Figure 1. For the proposed XML document, an interesting task may be the problem of mining frequent associations among people that appear as coauthors in the publications appearing in the XML document. In practice, we are interested in finding associations of the

```
XMINE RULE IN document("www.cs.atlantis.edu/research.xml") FOR
ROOT IN //People/*/Publications/* LET BODY := ROOT/Author,
    HEAD := ROOT/Author
EXTRACTING RULES WITH
 SUPPORT = 0.1 AND CONFIDENCE = 0.2
```

Fig. 3. The XMINE RULE statement for mining frequent associations among people that appear as coauthors in the publications appearing in the XML document of Figure 1

form: "{ *Wilson*} ⇒ {*Holmes*}" which states that, in the department, the papers which are authored by *Wilson* are also likely to have *Holmes* as author. The XMINE RULE statement for the mining problem introduced is reported in Figure 3.

2.3   Experimental Comparison of XML Mining Techniques

The approaches previously presented represent two different way to apply data mining techniques to XML data sources. On one side there are approaches that mines all the possible frequent patterns (e.g., [11]). On the other side there are approaches that mines only patterns that have a certain structure (e.g., [1]). In this paragraph we show some preliminary experiments we conducted over XML data using two implementations available for the two kind of approaches: X3 Miner [14] and XMine [15]. The purpose is not to assert which one of the two methods is better than the other, but to understand using real XML data, the distinguishing qualities of both methods.

As XML document to compare the two mining approaches, we used the *Digital Bibliography & Library Project* (DBLP) [17]. The DBLP XML document is a huge collection of bibliographic information on major computer science journals and proceedings. We made this chose because DBLP XML document is commonly considered the test bed for XML mining techniques. Both the implementations were not able to deal with the complete XML document within our test environment, so we build up a cut down version of the DBLP database containing more than 100, 000 publications from 2004 to 2006.

Figure 4 shows some results produced by X3 Miner applied to the DBLP XML document. For example, the first 3-item frequent pattern appears 256 times in the document and is composed by a root node `inproceedings` and two child nodes: `year` with value 2005 and `crossref` with value `conf/icdar/2005`. Actually that simply means that in the year 2005, have been published 256 conference articles at the conference ICDAR. It's easy to notice that X3 Miner tends to produce frequent tree patterns that represent the most frequent structures of the mined XML documents and hence, it may be used to mine documents where the structure is very rich and heterogeneous and where the information contained in the structure is relevant. The structure of DBLP XML document is very simple and homogeneous, hence patterns returned by X3 Miner are scarcely relevant. Results also suggests that X3 Miner can be applied to generate a simplified DTD or XML schema description of the XML document mined: X3 Miner returns the most frequent substructures and that this information, correctly interpreted, can be used to construct a DTD or XML schema of a XML document.

```
FREQUENT 1 ITEMS: booktitle[The Data Mining and Knowledge
Discovery Handbook]  - 68 volume[20]  - 1369 author[Wen Gao]  -
101 ...

FREQUENT 2 ITEMS: inproceedings author[Wen Gao]  - 87 article
journal[J. Symb. Comput.]  - 149 article year[2005]  - 16910 ...

FREQUENT 3 ITEMS: inproceedings year[2005] /
crossref[conf/icdar/2005]  - 256 article volume[21] /
journal[Bioinformatics]  - 608 inproceedings year[2004] /
booktitle[ICIP]  - 885 article volume[20] /
journal[Bioinformatics]  - 561 ...
```

Fig. 4. An example of frequent embedded subtrees extracted from the DBLP document using X3 Miner

```
<xmine_output variables="1265206" transactions="147266"
rules="5613">
  ...
  <RULE support="0.006" confidence="1.0">
    <BODY>
      <Item><booktitle>ICIP</booktitle></Item>
    </BODY>
    <HEAD>
      <Item><year>2004</year></Item>
    </HEAD>
  </RULE>
  ...
  <RULE support="0.0041" confidence="0.52">
    <BODY>
      <Item><journal>Bioinformatics</journal></Item>
    </BODY>
    <HEAD>
      <Item><volume>21</volume></Item>
    </HEAD>
  </RULE>
  ...
</xmine_output>
```

Fig. 5. An example of frequent association rules extracted from the DBLP document using XMine

While X3 Miner, given an XML document, returns all the frequent embedded XML subtrees, XMine needs a specific problem formulation to address the mining task. Considering the XML structure of the DBLP document, it seems natural to consider every publication as a "transaction" and each XML subtree of each publication as an "item".

Figure 5 reports some frequent association rules obtained by XMine from the DBLP XML document. It is easy to find correspondences between the rules extracted by XMine and the tree patterns obtained with X3 Miner. E.g., the tree pattern inproceedings year[2004] / booktitle[ICIP] (frequency 885) corresponds to the rule ⟨<booktitle>ICIP</booktitle>⟩ ⇒ ⟨<year>2004</year>⟩ (support 0.006 and confidence 1.0). An interesting thing is the added value given by the confidence of the mined association rule: in fact a 1.0 confidence

```
<xmine_output variables="348264" transactions="147266"
rules="418">
  <RULE support="0.0003" confidence="1.0">
    <BODY>
      <Item><author>Tomoya Enokido</author></Item>
    </BODY>
    <HEAD>
      <Item><author>Makoto Takizawa</author></Item>
    </HEAD>
  </RULE>
  ...
  <RULE support="0.00024" confidence="0.875">
    <BODY>
      <Item><author>Irith Pomeranz</author></Item>
    </BODY>
    <HEAD>
      <Item><author>Sudhakar M. Reddy</author></Item>
    </HEAD>
  </RULE>
  ...
</xmine_output>
```

Fig. 6. Association rules showing a co-author relationship extracted using XMine

means that the fragment `<booktitle>ICIP</booktitle>` is present in the XML document only with the fragment `<year>2004</year>`. We cannot find directly this information using X3 Miner.

Patterns extracted by the two miners do not reveal anything interesting. Considering the DBLP document, interesting patterns maybe, for example, patterns that shows associations between authors or authors and journals. To extract such kind of patterns, that may have a very low support – compared to not interesting patterns such the ones that may associate journals and years or volumes –, X3 Miner requires a huge computational effort, since it can not be instructed to mine only such patterns. Indeed, due to the low support of such pattern, we are not able to obtain them, since X3 Miner crashes before computing them. XMine, instead, can easily accomplish such a mining task and hence we can reduce the computational effort needed to extract such patterns. Furthermore, since X3 Miner extracts also the father of sibling nodes, it may happens that patterns having the same author are considered different because the father elements belong to a different type of publication (e.g., inproceedings or article). This may also results in not funding a pattern because the different patterns found are under the minimum required support. XMine can avoid such problem, if the root is expressed by mean of a generic XPath expression that includes every type of possible publication. XMine in this way easily returns some interesting patterns showing relationships between different authors that write often together (see Figure 6).

3   From Logs to Conceptual Logs

Web servers can collect large amount of information in their log files and in the log files of the databases they use. These logs usually contain basic information e.g.: name and IP of the remote host, date and time of the request, the request line exactly as it came from the client, etc. This information is usually represented in standard format e.g.: Common Log Format [18], Extended Log Format [19]. Common Log Format and Extended Log Format represent the standard log generated by common Web servers like Apache HTTP server [20]. Actually the most used log format on the Web is the Common Log Format (see Figure 7) or slight variation it.

As showed in Figure 7, a log generated by a Web server is a sequence of lines, where each line is composed by different fields – 9 in the case of the Common Log Format – with a precise role:

remotehost: This is the IP address of the client (remote host) which made the request to the server. The IP address reported here is not necessarily the address of the machine at which the user is sitting. If a proxy server exists between the user and the server, this address will be the address of the proxy, rather than the originating machine.

rfc931: The "hyphen" in the output indicates that the requested piece of information is not available. In this case, the information that is not available is the RFC 1413 identity of the client determined by identd on the client's machine.

authuser: This is the userid of the person requesting the document as determined by HTTP authentication. If the status code for the request (see below) is 401, then this value should not be trusted because the user is not yet authenticated. If the document is not password protected, this entry will be "-" just like the previous one.

[date]: The date and the time that the request was received.

"request": The request line from the client is given in double quotes. The request line contains a great deal of useful information. First, the method used by the client, second, the client requested the resource and third, the client used the protocol.

status: This is the status code that the server sends back to the client. This information is very valuable, because it reveals whether the request resulted in a successful response (codes beginning in 2), a redirection (codes beginning in 3), an error caused by the client (codes beginning in 4), or an error in the server (codes beginning in 5).

bytes: This entry indicates the size of the object returned to the client, not including the response headers. If no content was returned to the client, this value will be "-".

"referer": The "Referer" HTTP request header. This gives the site that the client reports having been referred from.

"user_agent": The User-Agent HTTP request header. This is the identifying information that the client browser reports about itself.

Recently, besides the de facto standards such previously introduced, the research of John Punin et al. proposed in [5] the specification of a new log format based on XML, the Log Markup Language (LOGML). LOGML files are obtained

```
xxx.xxx.xxx.xxx - - [11/Jun/2005...] "GET /index.jsp HTTP/1.0" 200 19454 "-" "Mozilla/4.0 (compatible; MSIE 6.0; Windows NT 5.0)"
yyy.yyy.yyy.yyy - - [11/Jun/2005...] "GET /page1.do?dau1.oid=321&UserCtxParam=0&GroupCtx..." 200 69687 "...2927" "Mozilla/4.0..."
xxx.xxx.xxx.xxx - - [11/Jun/2005...] "GET /DeiResources/sfondoNews.gif HTTP/1.0" 302 74 "...polimi.it/index.jsp" "Mozilla/4.0..."
yyy.yyy.yyy.yyy - - [11/Jun/2005...] "GET /upload/Matt..." 200 27374 "...&GroupCtxParam=0&ctx1=it&crc=371954722" "Mozilla/4.0..."
```

Fig. 7. An example of a web log formatted according the Common Log Format from the DEI Web site (`http://www.elet.polimi.it`): `remotehost rfc931 authuser [date]` ``request'' status bytes ``referer'' ``user_agent''. For privacy reasons, IP Addresses have been replaced with xs and ys sequences.

from standard Web logs (e.g., Common Log Format), and Web site maps expressed as XGMML, an XML language to describe graphs. LOGML generation was experimented for a simple static website. This log format is richer than the standards one as it includes the web site map, but still it does not contain any information about real page contents and as it is, it is not suitable for large data-driven Web sites (see Figure 8).

The LOGML is composed by two main subtrees: (i) `<graph>`, which contains the map describing the structure of the Web site and (ii) `<userSessions>`, that contains the information about user sessions. The `<graph>` is composed by `<node>` and `<edge>` elements. Nodes describes the different resources present in the Web site (e.g., html pages, gif, ...). The only interesting information provided for such resources is the title of the Web page. Edges describe the structure of the Web site connecting the different nodes according to links and the data contained in the source nodes. The subtree `<userSessions>`, is composed by `<userSession>` nodes describing each single user session by mean of edges. The XGMML graph describing the web sites structure can be reconstructed adopting some Web Structure Mining techniques like the one applied by Web spiders. It worths noticing that this task is quite expensive and it has to be performed again on each time the Web site structure changes.

The LOGML format, even if richer than common ones, still misses a rich description of the content of the navigated pages. Such description of content could be performed by extracting keywords or using other Web Content Mining techniques, again, augmenting the time and computational cost of generating such logs.

Intuitively, once data are available and their format can be easily processed, they can be trivially exploited to efficiently enrich Web logs. Web applications modeled and deployed using conceptual facilities can exploit the information contained in the conceptual schema of the Web application to enrich the Web logs. In particular we experimented the generation of conceptual Web logs within the WebML/WebRatio framework. WebML (Web Modeling Language) is a conceptual model for Web application design [8], which is an ingredient of a broader development methodology, supported by a CASE tool, named WebRatio [8,9]. WebML offers a set of visual primitives for defining conceptual schema that

```
<?xml version="1.0" encoding="UTF-8"?> <!-- LOGML File Generated
by webloggr 1.0 --> <logml start_date="12/Oct/2000:05:00:05"
end_date="12/Oct/2000:16:00:01">
  <graph directed="1">
    <node id="22" label="http://www.cs.rpi.edu/~puninj/XGMML/xgmml.dtd"
      hits="2" weight="2" ehits="1">
      <att name="title" value="No title"/>
      <att name="mime" value="text/html"/>
      <att name="size" value="6642"/>
      <att name="code" value="200"/>
    </node>
    <node id="21" label="http://www.cs.rpi.edu/~puninj/XGMML/draft-xgmml.html"
      hits="4" weight="4" ehits="3">
      <att name="title" value="XGMML (eXtensible Graph Markup and Modeling Language)
      1.0 Draft Specification"/>
      <att name="mime" value="text/html"/>
      <att name="size" value="126043"/>
      <att name="code" value="200"/>
    </node>
    ...
    <edge source="21" target="22" label="DTD" hits="1" weight="1"/>
    <edge source="3" target="10" label="Graph Gallery" hits="1" weight="1"/>
    <edge source="3" target="21" label="XGMML 1.0 Draft Specification" hits="1"
      weight="1"/>
    <edge source="3" target="11" label="XGMML 1.0 Draft Specification%0AUpdate"
      hits="1" weight="1"/>
    <edge source="2" target="8" label="Help File" hits="2" weight="2"/>
    <edge source="2" target="7" label=" ASHE's Slides" hits="2" weight="2"/>
    ...
  </graph>
  ...
  <userSessions count="2" max_edges="100" min_edges="2">
    <userSession name="yyy.yyy.yyy.yyy" ureferer="No_Referer"
      entry_page="http://www.cs.rpi.edu/~puninj/XGMML/"
      start_time="12/Oct/2000:12:50:11" access_count="4">
      <path count="3">
        <uedge source="3" target="10" utime="12/Oct/2000:12:50:12"/>
        <uedge source="3" target="21" utime="12/Oct/2000:12:51:41"/>
        <uedge source="21" target="22" utime="12/Oct/2000:12:52:02"/>
      </path>
    </userSession>
    <userSession name="xxx.xxx.xxx.xxx" ureferer="http://search.excite.com/search.gw?
      search=XHTML" entry_page="http://www.cs.rpi.edu/~puninj/TALK/head.html"
      start_time="12/Oct/2000:14:05:10" access_count="3">
      <path count="2">
        ...
      </path>
    </userSession>
  </userSessions>
</logml>
```

Fig. 8. An example of a LOGML: in the upper part the structure of the web site is reported and each page is tagged with is title, while in the lower part the user's session are reported

represent the organization of the application contents and of the hypertext interface. An example of conceptual model of a hypertext page, named *Teacher*, taken from the WebML-based hypertext schema of the http://www.elet.polimi.it application is reported in Figure 9. Besides having a visual representation, WebML primitives are also provided with an XML-based representation, to

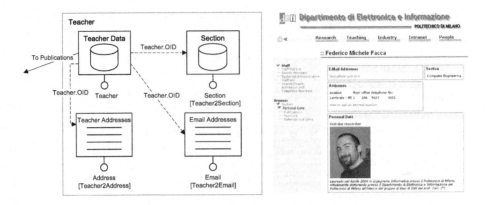

Fig. 9. The WebML schema of the page *Teacher* and its actual rendering

specify those additional properties that would not be conveniently expressible by a graphical notation. Figure 10 reports a simplified XML specification the previously introduced *Teacher* page. The page includes several content units. The first, a *data unit* publishes some attributes taken from a single instance of `Teacher`, which is an entity of the data schema. Moreover, from the *data unit* `dau1` a link originates, whose destination is a further unit (`dau132`) defined elsewhere in the application hypertext schema. A second *data unit* selects the instance to be published from the database according to a *selector condition*, specified over a relationship involving `Section`. Also, two *index units* are present in the page and publish lists of instances of entity `Address` and entity `Email`.

Webratio runtime environment for WebML applications logs not only the information collected normally from the Web servers but also the session identifier, e.g. `a_0YRnHNcly8`, allowing an easier reconstruction of user sessions.

The Webratio runtime provides also a *Runtime XML log* containing all the info about the processing of requested pages. The Runtime XML log includes all the events generated by the application runtime when serving a request page and populating its contents units. Each event is delimited in the XML log file by the `<event>` tag. Since each page request is managed by a specific thread, the events generated for a single page request are characterized by the same thread number. An `<event>` tag denotes either the request of an entire page, or the computation of an individual unit. It may contain further sub-tags:

- The tag `<message>` includes the event parameters. In case of content units population, it also includes the list of OIDs of the objects extracted from the data source.
- `<NDC>` stores the identifier of the conceptual element (page or unit) to which the event refers.

```
<PAGE auxiliary:split-subpages="yes" graphmetadata:go="page1_go"
 id="page1" landmark="no"
  localize="no" name="Teacher" presentation:page-layout="BasePage(+link)" secure="no">
   <CONTENTUNITS>
     <DATAUNIT entity="ent6" graphmetadata:go="dau1_go" id="dau1" inc-links="22"
       inc-links-from-dru="2" name="Teacher Data">
       <DISPLAYATTRIBUTE attribute="Name"/>
       <DISPLAYATTRIBUTE attribute="Surname"/>
       <DISPLAYATTRIBUTE attribute="Picture"/>
       <DISPLAYATTRIBUTE attribute="Curriculum"/>
       ...
       <LINK automaticCoupling="yes" graphmetadata:go="ln807_go" id="ln807"
         name="To Publications" newWindow="no" to="dau132" type="normal"/>
       ...
     </DATAUNIT>
     <DATAUNIT entity="Section" graphmetadata:go="dau175_go" id="dau175" inc-links="1"
       inc-links-from-dru="1" name="Section">
       <SELECTOR defaultPolicy="fill">
         <SELECTORCONDITION id="sel252" name="Teacher Section"
           predicate="in" relationship="Teacher2Section" type="required"/>
       </SELECTOR>
       <DISPLAYATTRIBUTE attribute="Name"/>
     </DATAUNIT>
     <INDEXUNIT distinct="no" entity="Address" graphmetadata:go="inu87_go"
       id="inu87" name="Addresses">
       <SELECTOR defaultPolicy="fill">
         <SELECTORCONDITION id="sel124" name="Teacher Addresses" predicate="in"
           relationship="Teacher2Address" type="required"/>
       </SELECTOR>
       <DISPLAYATTRIBUTE attribute="Location"/>
       <DISPLAYATTRIBUTE attribute="Floor"/>
       <DISPLAYATTRIBUTE attribute="Office"/>
       <DISPLAYATTRIBUTE attribute="Telephone"/>
       <DISPLAYATTRIBUTE attribute="Fax"/>
     </INDEXUNIT>
     <INDEXUNIT distinct="no" entity="Email" graphmetadata:go="inu43_go"
       id="inu43" name="Email Addresses">
       <SELECTOR defaultPolicy="fill">
         <SELECTORCONDITION id="sel11" name="Teacher Email Addresses" predicate="in"
           relationship="Teacher2Email" type="required"/>
       </SELECTOR>
       <DISPLAYATTRIBUTE attribute="Email"/>
     </INDEXUNIT>
   </CONTENTUNITS>
</PAGE>
```

Fig. 10. A portion of XML serialization of the WebML model for the page *Teacher*, of the DEI Web site (e.g., http://www.elet.polimi.it/people/facca)

For example:

```
<log4j:event category="/webapps/dei/log" index="72921"
timestamp="Fri, 11 June 2004 -
  02:02:01.140 AM" priority="DEBUG" thread="tcpConnection-80-4">
  <log4j:message>Continuing to serve request with id=-1503716237;
    remoteAddress=yyy.yyy.yyy.yyy; jSessionID=a_OYRnHNcly8;
    unitId=dau1; dataInstances=321;</log4j:message>
  <log4j:NDC>dau1</log4j:NDC>
</log4j:event>
```

is an event for a request to the *Teacher* page (see Figure 10). The event refers to the population of a *data unit* (dau1). Its <message> tag includes the unitID (dau1), the IP address and the client SessionID, and a value (321) representing the OID of the single database instance extracted for populating the data unit.

```
<ConceptualLog>
  <ConceptualSchema>
    ...
    <PAGE auxiliary:split-subpages="yes" graphmetadata:go="page1_go" id="page1" landmark="no"
    localize="no" name="Teacher" presentation:page-layout="BasePage(+link)" secure="no">
      <CONTENTUNITS>
        ...
      </CONTENTUNITS>
    </PAGE>
  </ConceptualSchema>
  <Log>
    <Session id="aFqaa3um9-1e">
      ...
    </Session>
    <Session id="a_0YRnHNcly8">
      <IPAddress>yyy.yyy.yyy.yyy</IPAddress>
      <HostName>###############</HostName>
      <Browser>Mozilla/4.0 (compatible; MSIE 6.0; Windows NT 5.0)</Browser>
      <StartTimestamp>1089583207000</StartTimestamp>
      <EndTimeStamp>1089583257000</EndTimeStamp>
      <Duration>50000</Duration>
      <Requests>
        <Request RequestId="0">
          ...
        </Request>
        <Request RequestId="1">
          <PageName>/page1.do</PageName>
          <Page SchemaRef="page1"/>
          <Referrer SchemaRef="page8"/>
          <EntryLink SchemaRef="ln51"/>
          <RequestType>GET</RequestType>
          <RequestURI>/page1.do?dau1.oid=321&UserCtxParam=0&GroupCtxParam=0
          &ctx1=it&crc=371954722</RequestURI>
          <Bytes>69687</Bytes>
          <Status>200 - OK</Status>
          <Referer>http://www.elet.polimi.it/page8.do?link=ln51.redirect&stu34.values=it
          &src7=rossi&src6=&alt2=page12&UserCtxParam=0&GroupCtxParam=0&ctx1=it</Referer>
          <RequestTimestamp>1089583208000</RequestTimestamp>
          <RequestTime>July 11, 2004 02:02:08 AM CEST</RequestTime>
          <ElapsedTime>1000</ElapsedTime>
          <PageUnits>
            <Unit>
              <Unit_Id SchemaRef="dau1"/>
              <DataInstance>321</DataInstance>
            </Unit>
            <Unit>
              <Unit_Id SchemaRef="dau175"/>
              <DataInstance>3</DataInstance>
            </Unit>
            <Unit>
              <Unit_Id SchemaRef="inu87"/>
              <DataInstance>700</DataInstance>
              <DataInstance>707</DataInstance>
            </Unit>
            <Unit>
              <Unit_Id SchemaRef="inu43"/>
              <DataInstance>402</DataInstance>
              <DataInstance>408</DataInstance>
            </Unit>
          </PageUnits>
        </Request>
        ...
      </Requests>
    </Session>
    ...
  </Log>
</ConceptualLog>
```

Fig. 11. A fragment of the Conceptual Log for a request to the *Teacher* page illustrated in Figure 10

The rich informations contained in Runtime XML log, application server log, and the Web application schema are easily exploited to generate an XML Conceptual log. A fragment of a XML Conceptual log is reported in Figure 11.

4   The Case Study Web Application

First experiments on Mining XML Conceptual logs were conducted on the WebML.org Web site (http://www.webml.org), the reference site for the WebML community, as reported in [7]. These first experiments were not enough to stress advantages of Conceptual Web logs, as the WebML site has a quite simple conceptual model and a limited number of daily visitors. To better prove the efficacy of our methodology, experiments on a more relevant Web application are needed.

Here we present the results achieved on mining Conceptual logs for the DEI Web site (http://www.elet.polimi.it): the Web site of the Computer Science Department of Politecnico di Milano. The DEI Web site contains more than 200 dynamic Web pages – modeled with WebML – gathering data from thousands of database's instances and more than 1,000 static Web pages belonging to the staff or to research groups. More than 2,000 people access the Web site everyday. These numbers are great enough to qualify the DEI Web application as a good field to deeply test the Conceptual Web log generation and to apply the XML Mining tasks. We generated Conceptual logs for 10 days from 11th June to 20th June 2005. The total original requests in the input Common Log Format files were more 1,500,000, while the final cleaned requests were about 350,000. The Table 1 presents some statistics about the generated Conceptual logs. The total number of users' sessions generated is 20,787 – not including robots' sessions – with an average of 17.3 requests per session.

The Web application we analyzed is mainly accessed by Italian students searching for information about teachers and courses. The public part of the application modeled with WebML is divided in five areas: (i) *Research* that provides information about research areas at the DEI; (ii) *Teaching* that proves info about examinations and other teaching related topics; (iii) *Intranet* that provides information to the DEI staff; (iv) *Companies* that provides information to the Companies who wants to collaborate with DEI; and (v) *Staff* that provides information about the DEI Staff. Figure 12(a) shows that the most

Table 1. Some statistics about the generated Conceptual logs for the DEI Web site

	Total	Daily Average	Daily Min	Daily Max
Original Requests	1,550,385	155,039	99,837	255,692
Filtered Requests	359,707	35,971	25,095	49,692
WebML Requests	74,697	7,470	3,785	10,518
DEI Personal Page Requests	183,200	18,320	12,726	25,600
DEI Generic Page Requests	62,518	6,252	4,245	8,144
Error Page Requests	38,073	3,807	2,086	5,283

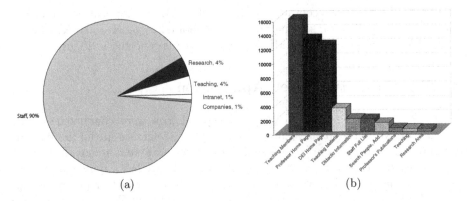

Fig. 12. Percentage of accesses to the different areas of the DEI Web site (a) and the top ten accessed pages (b)

accessed area is Staff (90%) that contains pages about Professors – the second most accessed page as shown in Figure 12(b) – and about Professors' teaching material – the forth most accessed page.

5   Results Analysis

In this section we report some of the interesting evidences we discovered analyzing the results of the different data mining tasks over the Conceptual Web logs generated for the DEI Web site.

Thanks to the richness of Conceptual logs, it is possible to extract knowledge at different levels of information abstraction. Some of the possible tasks are: (i) mining sequence over generic page sequences, i.e., considering only the sequences of page name accessed without referring to the data instances populating the page; (ii) mining sequences of data instanced pages; (iii) mining sequences of accessed entities and so on according to the selected fragments of the XML Conceptual log. The results published here regard task (i) and (ii). According to results obtained in Section 2 comparing different XML mining approaches, we adopted XMINE.

5.1   Accessing Professor Home Pages

As shown in Figure 12(b), most of the user interacting with the Web site are student navigating contents published by professors. This belief is supported not only by statistics but also by mining results. In fact, most of the discovered rules with the highest confidence, regards interactions with professors' pages. In particular, we identified the most common path to access content offered by professors (Figure 13): (i) the user access the "Search Members" page where he compiles the form to search a teacher, (ii) hence, from the list of results he chooses the pertinent one and accesses the "Teacher" page. The high confidence

```
<SequenceRule support="0.24770289123009573" confidence="0.8326326002587322">
    <AntecedentSequence>
        <ItemSet>
            <Item>
                <PageName>
                    <WebML_Page Id="page8" name="Search Members">
                        <SITEVIEW Id="sv1" name="Public">
                            <AREA Id="area5" name="Staff" landmark="yes"/>
                        </SITEVIEW>
                    </WebML_Page>
                </PageName>
            </Item>
        </ItemSet>
    </AntecedentSequence>
    <ConsequentSequence>
        <ItemSet>
            <Item>
                <PageName>formpage8</PageName>
            </Item>
        </ItemSet>
        <ItemSet>
            <Item>
                <PageName>
                    <WebML_Page Id="page1" name="Teacher">
                        <SITEVIEW Id="sv1" name="Public">
                            <AREA Id="area5" name="Staff" landmark="yes"/>
                        </SITEVIEW>
                    </WebML_Page>
                </PageName>
            </Item>
        </ItemSet>
    </ConsequentSequence>
</SequenceRule>
```

Fig. 13. Almost the 25% of the users' session contains a user request to the "Search Members" page followed by a search for a professor and a visit to a "Teacher" page. In particular 83% of the users that access the "Search Members" performs the form submission and accesses the "Teacher" page.

Fig. 14. An actual example of the user navigational path proposed by the rule depicted in Figure 13

of these rules is consistent with the fact that this path corresponds exactly to the Web application model and hence to the Web application designer objectives.

This is just one of the possible different paths that a user may navigate to access a "Teacher" page. The fact that this rule has much more higher support than rules regarding other paths to find and access professors' pages, supports the idea that this is the most efficient and effective path modeled within the Web application to satisfy such browsing goal.

Among the other frequent navigational paths performed by users to access "Teacher" pages, we notice a particular behavior for instances of teachers whose surname starts with 'A' letter. In such case we find that users prefer (i) getting the list of the whole Professors and then (ii) selecting the Professor from the first lines of the list and access his page. This is the only case where the "Staff Full List" page is used with a relevant support. This may suggest that a new list grouping professors by starting letter could be effective and useful.

Other derived rules show that students access available information in the "Teacher" page as expected. The highest support rules including the "Teacher" page as antecedent are: $Teacher \Rightarrow Available \quad Material$ (support 0.10 and confidence 0.23); $Teacher \Rightarrow Didactic \quad Information$ (support 0.08 and confidence 0.17) and $Teacher \Rightarrow Professor's \quad Publications$ (support 0.02 and confidence 0.05).

A discovered rule show that users that access the "Exams and Tests Results" page, then visit the "Member Search" page with a high confidence (0.22). Probably this behavior represents the fact that the "Exams and Tests Results" page is not often used by teachers to publish exams' results, and hence, when students do not find their test results in this page they look for the teacher page by accessing the "Member Search" page. This may suggest to include within the *Staff* area a page for the exams results of each teacher, trying to improve the way teachers publish their exams results and the way students can find them. We also obtain a number of sequential rules with a quite high confidence (around 0.15) showing a relationship between teachers. These relationships are not modeled within the Web application. Analyzing the teachers involved in these relationships, we discover that these navigational relationships correspond to real relationships between teachers. We find that the involved teachers are sharing the same course or teach courses belonging to the same class year. This may suggest to improve the Web application model, allowing to include automatically links to related teachers, and adding a page that groups teachers by class year so that students can directly access all the teachers relevant to their studies.

5.2   Misleading Link Names

One of the rule with highest support and confidence, shows that users that access the "Research" page – showing info about research sectors in which DEI is involved – then access the "Search Members" page – showing form to search for the Teaching Staff Members. This rule has a quite strong support (0.01, i.e. the sequence rule is reported in about 200 sessions over 20,000) and strong confidence (0.35, i.e. more than one third of the people that access the antecedent item of the rule then access the consequent item). To justify this navigational pattern we speculate that many users interpret the term "Research"[2] as *search for content within the DEI website*, and hence once they access the "Research" page and understand that is not what they are looking for, they move to the

---

[2] In Italian the term "Ricerca" is widely used with the meaning *search for something*.

page that seems most promising to support their task, in this case the "Search Members" page. This may suggest to better specify the concept underlying the term "Research" using a more rich periphrasis (e.g. "Scientific Research").

Another rule shows a frequent user behavior probably caused by inappropriate link naming: users that visit the "Technical-Administrative Staff List" page then often visit the "Search Members" page. Looking in the Web log we observe that users access the latter page most of the times immediately after the first one. Furthermore, most of the requests are generated starting from the DEI home page, where links to both the two pages are present. At the time we collected the Web logs, the links to the two pages had a quite similar name. The first link was named – in Italian – "Personale Non Docente" and the latter "Personale Docente". The two links were also one belove the other, increasing the chance of confusion for users[3]. Thus we speculate that the behavior showed by the rule may be caused by the similar names of the two links and by their adjacency.

6   Conclusion and Future Work

In this paper we presented the results of the extended mining experiments we performed over the Conceptual Web log generated from our Department Web site. The experiments supported our initial hypothesis that Conceptual Web logs allow for an easier task of analyzing mining results. Furthermore they easily allowed us to mine Web server logs at a different level of abstraction. The results reported evidence many problems in the current Web site and provide some directions that will be taken in account in upcoming restyling of our Department Web site. We are planning to add semantic annotation to the conceptual schema, as this should make easier the task of results analysis. This may also enable for a categorization of mined patterns according to concepts contained in the ontology used for annotation that may be exploited with clustering mining tasks.

Acknowledgments

The author wish to thank Pier Luca Lanzi and Maristella Matera for their precious suggestions and contributions that made this work possible.

References

1. Braga, D., Campi, A., Ceri, S., Klemettinen, M., Lanzi, P.: A tool for extracting xml association rules. In: Proceedings of ICTAI'02, 4-6 November, IEEE Computer Society, Los Alamitos (2002)
2. Cooley, R.: The use of web structure and content to identify subjectively interesting web usage patterns. ACM Trans. Inter. Tech. 3(2), 93–116 (2003)

---

[3] The Italian link name of the "Technical-Administrative Staff List" page has been recently updated to "Personale Tecnico-Amministrativo".

3. Stumme, G., Berendt, B., Hotho, A.: Usage mining for and on the semantic web. In: Next Generation Data Mining. Proc. NSF Workshop, Baltimore, November, 2002, pp. 77–86 (2002)
4. Heer, J., Chi, E.: Identification of web user traffic composition using multi-modal clustering and information scent. In: Proceedings of the Workshop on Web Mining, 2001 SIAM Conference on Data Mining (2001)
5. Punin, J.R., Krishnamoorthy, M.S., Zaki, M.J.: Logml: Log markup language for web usage mining. In: Kohavi, R., Masand, B., Spiliopoulou, M., Srivastava, J. (eds.) WEBKDD 2001 - Mining Web Log Data Across All Customers Touch Points. LNCS (LNAI), vol. 2356, Springer, Heidelberg (2002)
6. Facca, F.M., Lanzi, P.L.: Mining interesting knowledge from weblogs: a survey. Data & Knowledge Engineering 53(3), 225–241 (2005)
7. Fraternali, P., Lanzi, P.L., Matera, M., Maurino, A.: Model-driven web usage analysis for the evaluation of web application quality. J. Web Eng. 3(2), 124–152 (2004)
8. Ceri, S., Fraternali, P., Bongio, A., Brambilla, M., Comai, S., Matera, M.: Designing Data-Intensive Web Applications. Morgan Kaufmann, San Francisco, CA (2002)
9. Web Models: Webratio case tool (2006), http://www.webratio.com
10. Termier, A., Rousset, M.C., Sebag, M.: Treefinder: a first step towards xml data mining. In: Proceedings of the 2002 IEEE International Conference on Data Mining (ICDM 2002), pp. 450–457. IEEE Computer Society Press, Los Alamitos (2002)
11. Zaki, M.J., Aggarwal, C.C.: Xrules: an effective structural classifier for xml data. In: KDD '03: Proceedings of the ninth ACM SIGKDD international conference on Knowledge discovery and data mining, pp. 316–325. ACM Press, New York (2003)
12. Lian, W., Cheung, D.W., Mamoulis, N., Yiu, S.M.: An efficient and scalable algorithm for clustering xml documents by structure. IEEE Transactions on Knowledge and Data Engineering 16(1), 82–96 (2004)
13. Ahonen, H., Heinonen, O., Klemettinen, M., Verkamo, A.I.: Mining in the phrasal frontier. In: Komorowski, J., Żytkow, J.M. (eds.) PKDD 1997. LNCS, vol. 1263, pp. 343–350. Springer, Heidelberg (1997)
14. Tan, H., Dillon, T., Feng, L., Chang, E., Hadzic, F.: X3-Miner: Mining Patterns from XML Database. In: Proceedings of the 6th International Conference on Data Mining, Text Mining and their Business Applications, Skiathos, Greece (2005)
15. Facca, F.M.: Mining patterns from xml data: a structure-based approach. Master's thesis, Politecnico di Milano, Dipartimento di Elettronica e Informatica (2004)
16. Zaki, M.: Efficiently mining frequent trees in a forest. In: Hand, D., Keim, D., Ng, R. (eds.) Proceedings of the Eighth ACM SIGKDD International Conference on Knowledge Discovery and Data Mining (KDD-02), New York, ACM Press, pp. 71–80. ACM Press, New York (2002)
17. University of Trier, CS Department: DBLP - Digital Bibliography & Library Project (2006), http://dblp.uni-trier.de
18. Luotonen, A.: The common logfile format (1995), http://www.w3.org/pub/WWW/Daemon/User/Config/Logging.html
19. Hallam-Baker, P.M.: Extended log file format (1996), http://www.w3.org/TR/WD-logfile.html
20. The Apache Software Foundation: Apache http server project (2006), http://httpd.apache.org

Extracting and Using Attribute-Value Pairs from Product Descriptions on the Web

Katharina Probst[1], Rayid Ghani[1], Marko Krema[1], Andy Fano[1], and Yan Liu[2]

[1] Accenture Technology Labs, Chicago, IL, USA
[2] Carnegie Mellon University, Pittsburgh, PA, USA

Abstract. We describe an approach to extract attribute-value pairs from product descriptions in order to augment product databases by representing each product as a set of attribute-value pairs. Such a representation is useful for a variety of tasks where treating a product as a set of attribute-value pairs is more useful than as an atomic entity. We formulate the extraction task as a classification problem and use Naïve Bayes combined with a multi-view semi-supervised algorithm (co-EM). The extraction system requires very little initial user supervision: using unlabeled data, we automatically extract an initial seed list that serves as training data for the semi-supervised classification algorithm. The extracted attributes and values are then linked to form pairs using dependency information and co-location scores. We present promising results on product descriptions in two categories of sporting goods products. The extracted attribute-value pairs can be useful in a variety of applications, including product recommendations, product comparisons, and demand forecasting. In this paper, we describe one practical application of the extracted attribute-value pairs: a prototype of an Assortment Comparison Tool that allows retailers to compare their product assortments to those of their competitors. As the comparison is based on attributes and values, we can draw meaningful conclusions at a very fine-grained level. We present the details and research issues of such a tool, as well as the current state of our prototype.

1 Introduction

Retailers have been collecting a growing amount of sales data containing customer information and related transactions. These data warehouses also contain product information, but that information is often very sparse and limited. Specifically, most retailers treat their products as atomic entities with very few related attributes (typically brand, size, or color). Treating products as atomic entities hinders the effectiveness of many applications that businesses currently use transactional data for such as product recommendation, demand forecasting, assortment optimization, and assortment comparison. If a business could represent their products in terms of attributes and attribute values, all of the above applications could be improved significantly.

Suppose a sporting retailer wanted to forecast sales of a specific running shoe. Typically, they would look at the sales of the same product from the same time last year and adjust that based on new information. Now suppose that the shoe is described with the following attributes: *Lightweight mesh nylon material*, *Low Profile Sole*, *Standard lacing system*. Improved forecasting is possible if the retailer is able to describe the shoe

B. Berendt et al. (Eds.): WebMine 2006, LNAI 4737, pp. 41–60, 2007.

not only with a product number, but with a set of attribute-value pairs, such as *material: lightweight mesh nylon, sole: low profile, lacing system: standard*. This would enable the retailer to use data from other products having similar attributes. A similar argument can be made for building other applications listed above.

Many retailers have realized this recently and are trying to enrich their product databases with corresponding attributes and values for each product. In our discussions with retail experts, we found that in most cases, this is being done manually by looking at (natural language) product descriptions that are available in an internal database or on the web or by looking at the actual physical product packaging in the store. The work presented in this paper is motivated by the need to make the process of extracting attribute-value pairs from product descriptions more efficient and cheaper by developing an interactive tool that can help human experts with this task.

The task we tackle in this paper [4] requires a system that can process product descriptions and extract relevant attributes and values, and then form pairs by associating values with the attributes they describe. This can be accomplished by different means depending on the amount and type of information available. In many cases, product descriptions are only available in the form of unstructured text. This is the scenario for the work described here. The system described in this paper is able to extract attribute-value pairs from Web product descriptions with minimal human supervision. We describe the components of our system and show experimental results on a web catalog of sporting goods products.

As was said above, the extracted attribute-value pairs can be of use in a variety of practical applications. In this paper, we present one such practical application: an Assortment Comparison Tool that uses the automatically extracted attributes and values in order to compare a retailer's assortment to the assortment of the competitor. Comparison at the attribute-value (rather than the category or the individual product level) leads to much more fine-grained comparisons, allowing users to draw conclusions about differences in assortments.

2 Related Work

There has been a lot of research on extracting information from text documents on the Web but we are not aware of any system that addresses the same task as we are addressing in this paper. A related task that has received attention recently is that of extracting product features and their polarity from online user reviews.

Liu et al. [9] focus on extracting relevant product attributes, such as 'focus' in the domain of digital cameras. These attributes are extracted by use of a rule miner, and are restricted to noun phrases.The system then extracts polarized descriptors, e.g., 'good', 'too small', etc. Popescu and Etzioni [12] describe a similar approach: they first extract noun phrases as candidate attributes, and then compute the pointwise mutual information between the noun phrases and salient context patterns (such as *'scanner has'*). Similarly to Liu et al. [9], the extraction phase is followed by an opinion word extraction and polarity detection phase. Our work is similar in that a product is expressed as a vector of attributes. The difference is that our work focuses not only on attributes, but also on extracting values, and on associating the extracted attributes with the extracted values. Also, the attributes that are extracted from user reviews are often different (and described differently) than the attributes of the products that retailers would mention. For example, a review might mention 'photo quality' as an attribute but specifications of cameras would tend to use megapixels or the lens manufacturer in the specifications.

Information extraction with the goal of filling templates, e.g., [13,11], is related to the approach in this paper in that we extract certain parts of the text as relevant facts. It however also differs from such tasks in several ways, notably because we do not have a definitive list of 'template slots' available. Recent work in bootstrapping for information extraction using semi-supervised learning has focused on the task of named entity extraction [5,2,3], which is related to part of the work presented here (classifying the words/phrase as attributes or values or as neither).

3   The Attribute Extraction System

3.1   Overview of the Attribute Extraction System

Our extraction system consists of five modules: 1) Data Collection, 2) Seed Generation, 3) Attribute-Value Entity Extraction, 4) Attribute-Value Pair Relationship Extraction, and 5) User Interaction. The modular design allows us to break the problem into smaller steps, each of which can be addressed by various approaches. In this paper, we have chosen one specific approach for each phase. We only focus on tasks 1-4 in this paper, where task 5 is largely future work that we however consider very important.

3.2   Data

The data required for extracting product attributes and values can come from a variety of sources such as an internal product database or from the retailer website. To experiment with our extraction algorithms, we crawled the web site of a sporting goods retailer (www.dickssportinggoods.com), concentrating on the domains of tennis and football. Sporting goods is an interesting and relatively challenging domain because unlike electronics, the attributes are not easy and straightforward to detect. For example, a camera has a relatively well-defined list of attributes (*resolution, zoom, memory-type*, etc.). In contrast, a baseball bat would have some typical attributes such as brand, length, and material as well as others that might be harder to identify as attributes and values (*aerodynamic construction, curved hitting surface*, etc).

The scraping process resulted in a set of product descriptions where each product is described by a list of phrases, which we use as training data. Some examples of entries in these lists are *1 tape cutter, 4 rolls of white athletic tape, Cutout midfoot, Extended Torsion bar, Synthetic leather upper, Audio/Video Input Jack, Play Dry technology offers moisture management and wicking properties, Vulcanized latex outsole construction is lightweight and flexible.*

It can be seen from these examples that the entries are not often full sentences. This makes the extraction task more difficult, because most of the phrases contain a number of modifiers. There is often no definitive answer as to what the extracted attribute-value pair should be, even for humans inspecting the data. For instance, should the system extract *cutter* as an attribute with two separate values, *1* and *tape*, or should it rather extract *tape cutter* as an attribute and *1* as a value? To answer this question, it is important to keep in mind the goal of the system to express each product as a vector of attribute-value pairs, so as to compare between products. Therefore, it is more important that the system is consistent than which of the valid answers it gives.

3.3 Pre-processsing

The product descriptions collected by the web crawler are first tagged with parts of speech (POS) using the Brill tagger and stemmed with the Porter stemmer. We also replace all numbers with the unique token *#number#* and all measures (e.g., *liter, kg*) by the unique token *#uom#*. Additionally, we compute several correlation scores (Yule's Q statistic, pointwise mutual information, and information gain) between all pairs of words and recognized one as a phrase if all of its correlation scores exceed certain thresholds. In the experiments reported in section 4 below, we set thresholds for the correlation scores that in our experience yield robust results: for Yule's Q statistic, we used 0.98500, for mutual information, we used a threshold of 5.5, and for information gain, we used 0.002.

3.4 Seed Generation

Once the data is collected and processed, the next step is to provide labeled seeds for the learning algorithms to learn from. The extraction algorithm is seeded in two ways: with a list of known attributes and values, as well as with an unsupervised, automated algorithm that extracts a set of seed pairs from the unlabeled data. Both of these seeding mechanisms are designed to facilitate scaling to other domains.

Generic and domain-specific lists as labeled seeds. We use a very small amount of labeled data in the form of generic and domain-specific lists. The generic value lists were easily available on the web and are fairly domain-independent. We use lists of colors, materials, countries, and units of measure. In addition, we use a list of domain-specific (in our case, sports) values and attributes consisting of sports teams (such as *Pittsburgh Steelers*).

These seeds are supplemented by automatically extracted attribute-value seed pairs, as described in the following section. In other words, aside from easily replaceable generic and domain-specific lists, the system works in an unsupervised fashion.

Unsupervised Seed Generation. Typically, supervised and semi-supervised learning algorithms required labeled data. We developed an unsupervised algorithm that is able to generate labeled data (seeds), eliminating the need to manually provide labeled data. Our unsupervised seed generation method extracts a small number of attribute-value pairs from the unlabeled data that serve as labeled data for classification. We use correlation scores to find candidates, and make use of POS tags by excluding certain words from being candidates for extraction.

Extracting attribute-value pairs is related to the problem of phrase recognition in that both methods aim at extracting pairs of highly correlated words. There are however differences between the two problems. Consider the following two sets of phrases: *back pockets, front pockets, zip pockets* as compared to *Pittsburgh Steelers, Chicago Bears.*

The first list contains an example of an attribute with several possible values. The second list contains phrases that are not attribute-value pairs. The biggest difference between the two lists is that attributes generally have more than one possible value, as in the above example. We exploit this observation to automatically extract high-quality seeds by defining a modified mutual information metric as follows.

We consider all bigrams $w_i w_{i+1}$ as candidates for pairs, where w_i is a candidate value, and w_{i+1} is a candidate attribute. Although the modifying value does not always occur (directly) before its attribute, this heuristic allows us to extract seeds with high precision. Suppose word w (in position $i+1$) occurs with n unique words $w_{1...n}$ in position i. We rank the words $w_{1...n}$ by their conditional probability $p(w_j|w), w_j \in w_{1...n}$, where the word w_j with the highest conditional probability is ranked highest.

The words w_j that have the highest conditional probability are candidates for values for the candidate attribute w. Clearly, however, not all words are good candidate attributes. We observed that attributes generally have more than one value and typically do not occur with a wide range of words. For example, frequent words such as *the* occur with many different words. This is indicated by their conditional probability mass being distributed over a large number of words. We are interested in cases where few words account for a high proportion of the probability mass. For example, both *Steelers* and *on* will not be good candidates for being attributes. *Steelers* only occurs after *Pittsburgh* so all of the conditional probability mass will be distributed on one value whereas *on* occurs with many words with the mass distributed over too many values. This goal can be accomplished in two phases: in the first phase, we retain enough words w_j to account for a part $z, 0 < z < 1$, of the conditional probability mass $\sum_{j=1}^{k} p(w_j|w)$. In the experiments reported here, z was set to 0.5.

In the second phase, we compute the *cumulative* modified mutual information for all candidate attribute-value pairs. We again consider the perspective of the candidate attribute. If there are a few words that together have a high mutual information with the candidate attribute, then we are likely to have found an attribute and (some of) its values. We define the cumulative modified mutual information as follows:

Let $p(w, w_{1...k}) = \sum_{j=1}^{k} p(w, w_j)$. Then

$$cmi(w_{1...k}; w) = \log \frac{p(w, w_{1...k})}{(\lambda * \sum_{j=1}^{k} p(w_j)) * ((\lambda - 1) * p(w))}$$

λ is a user-specified parameter, where $0 < \lambda < 1$. We have experimented with several values, and have found that setting $\lambda \approx 1$ yields robust results. Setting $\lambda \approx 0$ implies that a candidate pair is not penalized for the word w being frequent, as long as few words cover most of its conditional probability mass. Table 1 lists several examples of extracted attribute-value pairs.

As we can observe from the table, our unsupervised seed generation algorithm captures the intuition we described earlier and extracts high-quality seeds for training the system. We expect to refine this method in the future. Currently, not all extracted pairs are actual attribute-value pairs. One typical example of an extracted incorrect pair are first name - last name pairs, e.g., *Smith* is extracted as an attribute as it occurs as part of many phrases and fulfills our criteria (*Joe Smith, Mike Smith*, etc.) after many first names. Other examples of incorrectly extracted attribute-value pairs include '*more*

Table 1. Automatically extracted seed attribute-value pairs

value	attribute
carrying, storage	case
main, racquet	compartment
ball, welt, side-seam, key	pocket
coat, durable	steel

(attribute) – *much* (value)' and '*more* (attribute) – *achieve* (value)'. However, some of the incorrectly extracted examples are rare enough that they do not have much impact on subsequent steps. The current metric accomplishes about 65% accuracy in the tennis category and about 68% accuracy in the football category. We have experimented with manually correcting the seeds by eliminating all those that were incorrect. This did not result in any improvement of the final performance of the overall system, leading us to conclude that our algorithm is robust to noise and is able to deal with noisy seeds.

For exploratory purposes, we also experimented with labeled training data: the data that we used in our experiments exhibits structural patterns that clearly indicate attribute-value pairs separated by colons, e.g., *length: 5 inches*. The results obtained with the automatically extracted pairs are comparable to the ones obtained when the given attribute-value pairs were used. This is probably because the labeled pairs were not very useful training examples for classification because it was hard for the learning algorithms to generalize from them. In the case of comparable results we prefer our approach of minimal supervision and minimal reliance on structural web site patterns, because it enables higher domain and data set independence.

3.5   Attribute and Value Extraction

After generating initial seeds, the next step is to use the seeds as labeled training data to extract attributes and values from the unlabeled data. We formulate the extraction as a classification problem where each word or phrase can be classified as an attribute or a value (or as neither). We treat it as a supervised learning problem and use Naïve Bayes as our first approach. The initial seed training data is generated as described in the previous section and serves as labeled data which Naïve Bayes uses to train a classifier. Since our goal is to create a system that minimizes human effort required to train the system, we use semi-supervised learning to improve the performance of Naïve Bayes by exploiting large amounts of unlabeled data available for free on the Web. Gathering product descriptions (from retail websites) is a relatively cheap process using simple web crawlers. The expensive part is labeling the words in the descriptions as attributes or values. We augment the initial seeds (labeled data) with the all the unlabeled product descriptions collected in the data collection phase and use semi-supervised learning (co-EM [10] with Naïve Bayes) to improve attribute-value extraction performance. The classification algorithm is described in the sections below.

Initial labeling. The initial labeling of data items (words or phrases) is based on whether they match the labeled data. We define four classes to classify words into:

unassigned, attribute, value, or neither. The initial label for each word defaults to *unassigned* and is changed to the label of any labeled data that it matches or to *neither* if it is a stopword.

Naïve Bayes Classification. The labeled words are then used as training data for Naïve Bayes that classifies each word or phrase in the unlabeled data as an attribute, a value, or neither. The features used for classification are the words of each unlabeled data item, plus the surrounding 8 words and their corresponding parts of speech. With this feature set, we capture not only each word, but also its context as well as the parts of speech in its context. This is similar to earlier work in extracting named entities using labeled and unlabeled data [3].

co-EM for Attribute Extraction. Since labeling attributes and values is an expensive process, we use the semi-supervised learning setting by combining small amounts of labeled data with large amounts of unlabeled data. We use the multi-view or co-training [1] setting, where each example can be described by multiple views (e.g., the word itself and the context in which it occurs). The specific algorithm we use is co-EM [10]. Co-EM with Naïve Bayes has been applied to classification, e.g., by [10], but so far as we are aware, not in the context of information extraction. The separation into feature sets we use is that of the word to be classified and the context in which it occurs. Each word is expressed in *view1* by the stemmed word itself, plus the part of speech as assigned by the Brill tagger. The *view2* for this data item is a context of window size 8, i.e. up to 4 words (plus parts of speech) before and up to 4 words (plus parts of speech) after the word or phrase in *view1*. If the context around a *view1* data item is less than 8 words long, we simply limit to the context to what is available.

co-EM Algorithm. co-EM proceeds by initializing the *view1* classifier using the labeled data only. Then this classifier is used to probabilistically label all the unlabeled data. The context (*view2*) classifier is then trained using the original labeled data plus the unlabeled data with the labels provided by the *view1* classifier. Similarly, the *view2* classifier then relabels the data for use by the *view1* classifier, and this process iterates for a number of iterations or until the classifiers converge.

Each iteration consists of collecting evidence for each data item from all the data items in the other view that it occurs with. For example, if a *view2* data item $view2_k$ occurs with (i.e., in the context of) *view1* data items $view1_{i1}$ and $view1_{i2}$, then the probability distribution for $view2_k$ is the averaged distribution of the probabilities currently assigned to $view1_{i1}$ and $view1_{i2}$, weighted by the number of times $view2_k$ appears together with $view1_{i1}$ and $view1_{i2}$, respectively, as well as by the class priors.Our goal is to label unlabeled training examples that are attributes or values, and leave the others unlabeled. Co-EM can be summarized by the following steps: 1) Initialize based on labeled data (see above). 2) Use $view1$ to label $view2$. 3) Use $view2$ to label $view1$. 4) Repeat for steps 2 and 3 n iterations. 5) Assign final labels to words using the predictions from both views.

Estimating class priors. When estimating class priors for labeling a view, the class priors are estimated from the respective other view's probability distributions. As each data item is associated with a set of data items from the other view with which it co-occurs, together with a count of how many times the two data items co-occurred, we could gather the class prior information by traversing through each data item and weighing the probability distributions from the aligned data elements by the co-occurrence counts.

Conceptually, however, it is easier to think of the class priors as simply obtained from the training data's current distribution in the other view. In other words, when labeling $view2$ from $view1$, the class priors for the Naïve Bayes classifier are computed only on $view1$, without reference to the $view2$ data items. The resulting probability distributions from these two approaches are the same.

The class probabilities are thus estimated as follows:

$$P(c_k) = \frac{1 + \sum_i^{n_1} cnt(view1_i) * P(c_k|view1_i)}{numclasses + \sum_i^{n_1} cnt(view1_i)}$$

Estimating word probabilities. As with class priors, word probabilities from $view1$ are used as training data for $view2$. For example, if a $view1$ element has a probability distribution of $p(value) = 0.5$ and $p(attribute) = 0.5$, then the data element is counted as a value example with weight 0.5, but also as an attribute example with weight 0.5.

For all words $view2_j$, estimate the new probability for each class $c_k, 1 \leq k \leq 4$, from all words $view1_i, 1 \leq i \leq n_1$. In practice, the algorithm considers only those $view2_j$ items whose cooccurrence count with $view1_i$ is greater than zero.

$$P(view2_j|c_k) = \frac{1 + \sum_{i=1}^{n_1} cooc(view1_i, view2_j) * P(c_k|view1_i)}{n_2 + \sum_{i=1}^{n_1} cooc(view1_i, view2_j)}$$

$$P(view1_i|c_k) = \frac{1 + \sum_{j=1}^{n_2} cooc(view1_i, view2_j) * P(c_k|view2_j)}{n_1 + \sum_{j=1}^{n_2} cooc(view1_i, view2_j)}$$

Labeling unlabeled examples. In each iteration, we want to use the computed class and word probabilities to label unlabeled data items in the respective other view. This can be done as follows:

$$P(c_k|view2_i) \propto P(c_k) * P(view2_i|c_k)$$

if $view2_i$ does *not* match the labeled training data.

After computing the probabilities for all classes, we must renormalize:

$$P(c_k|view2_j) = \frac{P(c_k|view2_j)}{\sum_{k=1}^{numclasses} P(c_k|view2_j)}$$

However, if $view2_i$ matches the labeled training data,

$$P(c_k|view2_i) = Initial Labeling.$$

$$P(c_k|view1_i) \propto P(c_k) * P(view1_i|c_k)$$

if $view1_i$ does *not* match the labeled training data. As in the case of $view2$, we will need to renormalize after computing the probabilities for each class. Also as above, if $view1_i$ matches the labeled training data,

$$P(c_k|view1_i) = InitialLabeling.$$

Assigning co-EM probabilities to $\langle view1_i, view2_j \rangle$ **pairs.** After co-EM is run for a pre-specified number of iterations, we assign final co-EM probability distributions to all $\langle view1_i, view2_j \rangle$ pairs as follows:

$$P(c_k|\langle view1_i, view2_j \rangle) = \frac{P(c_k|view1_i) + P(c_k|view2_j)}{2}$$

Final labels are assigned to words and phrases by averaging the predictions of each view's classifier. It should be noted that words that are tagged as attributes or values are not necessarily extracted as part of an attribute-value pair in the next phase. They will only be extracted if they form part of a pair, or if they occur frequently enough by themselves or as part of a longer phrase. The next section will describe this in greater detail.

3.6 Finding Attribute-Value Pairs

After the classification algorithm has assigned a (probabilistic) label to all unlabeled words, a final important step remains: using these labels to tag attributes and values in the actual product descriptions, i.e., in the original data, and finding correspondences between words or phrases tagged as attributes and values. The classification phase assigns a probability distribution over all the labels to each word (or phrase). This is not enough, because aside from n-grams that are obviously phrases, some subsequent words that are tagged with the same label should be *merged* to form an attribute or value phrase. Additionally, the system must establish *links* between attributes (or attribute phrases) and their corresponding values (or value phrases), so as to form attribute-value pairs. Some unlabeled data items contain more than one attribute and more than one value, so that it is important to find the correct associations between them. We accomplish merging and linking in an interleaved fashion, using the following steps:

- **1:** Link attributes and values if they match a seed pair.
- **2:** Merge words of the same label into phrases if their correlation scores exceed a threshold.
- **3:** Link attribute and value phrases based on directed dependencies as given by a dependency parser [8]: attribute phrases and value phrases can form a pair if there is a governor-dependent relationship between them.
- **4:** Link attribute and value phrases if they exceed a correlation score threshold: unassigned attribute phrases are linked with value phrases if their words exceed a correlation threshold.
- **5:** Link attribute and value phrases based on proximity: unassigned attribute phrases are linked with value phrases if if they are adjacent.

- **6:** Adding known, but not overt, attributes: material, country, and/or color.
- **7:** Extract binary attributes, i.e., attributes without values, if they appear frequently or if the unlabeled data item consists of only one word.

In the process of establishing attribute-value pairs, we exclude words of certain parts of speech, namely most closed-class items such as prepositions and conjunctions.

In step 6,the system 'extracts' information that is not explicit in the data. The attributes (*country*, *color*, and/or *material*) are added to any existing attribute words for this value if the value is on the list of known countries, colors, and/or materials. Assigning attributes from known lists is an initial approach to extracting non-explicit attributes. In the future, we will explore this issue in greater details.

Even after all the above pair identification steps, some attribute or value phrases can remain unaffiliated. Some of them are extracted noise, and should not be output. Others are valid attributes with binary values. For instance, the data item *Imported* is a valid attribute with two possible values: *true* or *false*, where the value is simply assigned by the absence or presence of the attribute. We extract only those attributes that are single word data items and those attributes that occur frequently in the data as a phrase.

4   Evaluation

4.1   Attribute-Value Pairs Extraction

We present evaluation results for experiments performed on tennis and football categories. The tennis category contains 3194 unlabeled data items (i.e., individual phrases from the bulleted list of product descriptions), the football category 72825 items. Automated seed extraction resulted in 169 attribute-value pairs for the tennis category and 180 pairs for football. Table 2 shows a sample list of extracted attribute-value pairs (i.e., the output of the full system), and the phrases that they were extracted from.

We ran our system in the following three settings to gauge the effectiveness of each component: 1) only using the automatically generated seeds and the generic lists

Table 2. Examples of extracted pairs for system run with co-EM

Full Example	Attribute	Value
1 1/2-inch polycotton blend tape	polycotton blend tape	1 1/2-inch
1 roll underwrap	underwrap	1 roll
1 tape cutter	tape cutter	1
Extended Torsion bar	bar	Torsion
Synthetic leather upper	#material# upper	leather
Metal ghillies	#material# ghillies	Metal
adiWear tough rubber outsole	rubber outsole	adiWear tough
Imported	Imported	#true#
Dual-density padding with Kinetofoam	padding	Dual-density
Contains 2 BIOflex concentric circle magnet	BIOflex concentric circle magnet	2
93% nylon, 7% spandex	#material#	93% nylon 7% spandex
10-second start-up time delay	start-up time delay	10-second

('Seeds' in the tables), 2) with the baseline Naïve Bayes classifier ('NB'), and 3) co-EM with Naïve Bayes ('coEM'). To make the experiments comparable, we do not vary pre-processing or seed generation, and keep the pair identification steps constant as well.

The evaluation of this task is not straightforward. The main problem is that people often do not agree on what the 'correct' attribute-value pair should be. Consider the example *Audio/JPEG navigation menu*. This phrase can be expressed as an attribute-value pair in multiple ways:

Possible Attribute	Possible Value
navigation menu	*Audio/JPEG*
menu	*Audio/JPEG navigation*
Audio/JPEG navigation menu	*#true#*

In the last case, the entire phrase is considered a binary attribute. All three pairs are both possibly useful attribute-value pairs. The implication is that a human annotator will make one decision, while the system may make a different decision (with both of them being consistent). For this reason, we give partial credit to an automatically extracted attribute-value pair, even if it does not completely match the human annotation. In some cases, an extracted pair deserves only partial credit, while in other cases, the automatically extracted pair is an equally valid attribute-valid pair.

For each of the metrics, we report *type* and *token* performance. Type performance (at the data item level, i.e., at the level of individual product description phrases) refers to performance for unique examples (each example contributes the same regardless of frequency). The data sets contain a number of duplicates, as many attributes apply to more than one product. Token performance refers to performance including duplicates, therefore emphasizing those examples that occur more frequently than others.

Precision. To measure precision, we evaluate how many automatically extracted pairs match manual pairs completely, partially, or not at all. The percentage of pairs that are fully *or* partially correct is useful as a metric especially in the context of human post-processing: partially correct pairs are corrected faster than completely incorrect pairs. Tables 3 and 4 list results for this metric for both categories, and for both *type* and *token* evaluations.

The results show that all three systems achieve very high performance for partially correct pairs. As expected, seed generation alone achieves higher accuracy than the system achieves when using unlabeled, and thus noisy, data. As we will see in the

Table 3. *Type* (left) and *Token* (right) Precision for *Tennis* Category

	Seeds	NB	coEM		Seeds	NB	coEM
# corr pairs	14	20	50	# corr pairs	252	264	316
# part corr pairs	54	73	132	# part corr pairs	202	247	378
% fully correct	**20.29**	**21.28**	**26.60**	**% fully correct**	**54.90**	**51.16**	**44.44**
% full or part correct	**98.56**	**98.94**	**96.81**	**% full or part correct**	**98.91**	**99.03**	**97.60**
% incorrect	**1.44**	**1.06**	**3.19**	**% incorrect**	**1.08**	**0.97**	**2.39**

Table 4. *Type* (left) and *Token* (right) Precision for *Football* Category

	Seeds	NB	coEM		Seeds	NB	coEM
# corr pairs	12	18	39	# corr pairs	4704	5055	6639
# part corr pairs	63	95	159	# part corr pairs	8398	10256	13435
% fully correct	**15.38**	**14.44**	**17.65**	**% fully correct**	**35.39**	**31.85**	**32.04**
% full or part correct	**96.15**	**90.40**	**89.59**	**% part or full correct**	**98.56**	**96.48**	**96.88**
% incorrect	**3.85**	**9.60**	**10.41**	**% incorrect**	**1.44**	**3.52**	**3.12**

following section, however, the decrease in precision when co-EM is used is more than offset by a large increase in recall.

Recall. Whenever the system extracts a partially correct pair for an example that is also given by the human annotator, the pair is considered recalled. The results for this metric can be found in tables 5 and 6. Unlike for precision, the recall differs greatly between system settings. More specifically, co-EM aids in recalling a much larger number of pairs, whereas seed generation and Naïve Bayes result in relatively poor recall performance.

Table 5. *Type* (left) and *Token* (right) Recall for *Tennis* Category

	Seeds	NB	coEM		Seeds	NB	coEM
# recalled	66	87	167	# recalled	451	502	668
% recalled	**27.62**	**36.40**	**69.87**	**% recalled**	**51.25**	**57.05**	**75.91**

Table 6. *Type* (left) and *Token* (right) Recall for *Football* Category

	Seeds	NB	coEM		Seeds	NB	coEM
# recalled	68	98	164	# recalled	12629	14617	17868
% recalled	**32.69**	**47.12**	**78.85**	**% recalled**	**39.21**	**45.38**	**55.48**

Word-based Label-independent Precision and Recall. Often there is partial overlap between an automatically extracted pair and a pair given by a human annotator. Sometimes both pairs are equally valid, and sometimes the automatically pair is useful even if it is not completely correct, because it can easily be corrected by a human annotator. This gives us an idea of how well the system can predict that a word should be part of a pair, even though it may confuse whether the word should be tagged as an attribute or a value. We define precision, recall, and F1 in the standard way. We also measure the amount of 'confusion', i.e., how often a (human-tagged) value word was automatically labeled as an attribute or vice versa. For this metric, the performance of co-EM and NB are quite comparable, but both greatly outperform seed generation only in recall.

As was discussed in the seed extraction section, we experimented also with correcting the automatically extracted seeds and running our system with the corrected seeds. This experiment was run only for tennis with co-EM. The result was no significant change in performance. This leads us to conclude that our algorithm is quite robust to noise. It also leads us to the conclusion that the time of a human annotator is likely better spent correcting the final output of the system rather than the input seeds. Correcting the input seeds does not necessarily lead to improved performance, whereas correcting complete output pairs is likely to do so. We will explore this issue further in the context of the active learning phase in our system.

Precision Results for Most Frequent Data Items. As the training data contains many duplicates, it is more important to extract correct pairs for the most frequent pairs than for the less frequent ones. In this section, we report precision results for the most frequently data items. This is done by sorting the training data by frequency, and then manually inspecting the pairs that the system extracted for the most frequent 300 data items. This was done only for the system run that includes co-EM classification. We report precision results for the two categories (*tennis* and *football*) in two ways: first, we do a simple evaluation of each unique data item. Then we weight the precision results by the frequency of each sentence. In order to be consistent with the results from the previous section, we define five categories that capture very similar information to the information provided above. The five categories contain *fully correct* and *incorrect*. Another category is *Flip to correct*, meaning that the extracted pair would be *fully* correct if attribute and value were flipped. *Flip to partially correct* refers to pairs that would be *partially* correct if attribute and value were flipped. Finally, we define *partially correct* as before. Table 7 shows the results.

Table 7. *Non-weighted* and *Weighted* Precision Results for *Tennis* and *Football* Categories. 'T' stands for *tennis*, 'F' is *football*, 'nW' *non-weighted*, and 'W' is *weighted*

	T nW	T W	F nW	F W
% fully correct	51.25	55.89	51.90	60.01
% flip to correct	12.08	20.14	9.62	10.25
% flip to partially correct	2.92	1.75	0.87	2.14
% partially correct	32.92	21.74	35.27	25.98

Discussion. The results show that we can learn product attribute-value pairs in a largely unsupervised fashion with encouraging results. One conclusion is that there is some confusion over which label an extracted word or phrase should have. This is consistent with human disagreement over the labels. Confusion levels increase when co-EM is added to the system, indicating that there were not enough seeds to train a strong classifier to differentiate between attributes and values. Future work will include user-specified lists that can serve as attribute seeds. Such labeled examples can be provided as part of an interactive step or before learning takes place, as is done currently.

The baseline Naïve Bayes algorithm also outperforms the seed generation algorithm (not using co-EM) in recall. This is not surprising, as the seeds are used as labeled training data for Naïve Bayes, which in turn labels additional examples that cannot be labeled by the seeds only. It does, however, not match the recall performance of co-EM, and only outperforms co-EM slightly in terms of precision for *tennis*, but not so for *football*.

Evaluating precision on the most frequent data items yields similar results. We show that there are few incorrect pairs, and we show that especially if we weight by the frequency, the number of completely correct examples is encouragingly high. Furthermore, a fair number of examples can become completely correct if flipped. In the future, we will investigate techniques to detect pairs that should be flipped, which could lead to improved precision. Finally, we can conclude that the results are consistent for both categories, making a strong case for the scalability of the system to other domains.

4.2   Exploiting Extracted Attribute-Value Pairs: A Practical Application

In the previous sections, we described a system that is able to extract attributes and values from product descriptions. Treating products as a set of attribute and values instead of atomic entities enables a variety of business applications. One such application is an Assortment Comparison Tool, which we describe in this section.

Today, when retailers compare their own offerings with those of competitors, the analysis is not principled. While retailers may be aware of many of the product categories or even individual products that are also carried by their competitors, they do not have a clear understanding of how their assortment differs from their competitors' assortments. The Assortment Comparison Tool that we present in this section will allow retailers to explore, in a principled way, their own as well as their competitors' assortments and draw conclusions from the comparison.

Retailers currently compare assortments either at the category level or at the individual product (SKU) level. At the category level, two retailers can have comparable assortments if they carry products in the same category. That is certainly not enough to achieve a detailed understanding of the differences between assortments. At the individual product level, products are only comparable if they can be identified as exactly the same. What is missing is a *similarity* metric between two different products. Expressing each product as a vector of attribute-value pairs will allow for such a similarity metric.

More specifically, the Assortment Comparison Tool compares products on a *attribute-value* level, offering the opportunity for very fine-grained comparison. For instance, we notice that a competitor offers many more cameras with 8 megapixels. If in general 8-megapixel cameras sell well, the system may suggest to increase assortment in this space. In the following, we delve into more details of the assortment comparison tool, its implementation and research challenges, as well as its current state.

Products are described by attribute-value pairs using the attribute extraction system described above. In order to compare assortments, a series of mappings between categories and products will need to take place. More specifically, each retailer tends to have their own specific product hierarchy, and hierarchy matching must be done in order to compare. Similarly, the extracted attributes and values are likely to be consistent for products within a retailer but can differ across retailers, requiring a mapping across retailers to draw comparisons.

The first step in comparing product assortments, mapping product hierarchies, is a non-trivial problem: the product hierarchies used by retailers differ more than one might expect. The categories are often ad hoc and the decision to create a particular subcat-

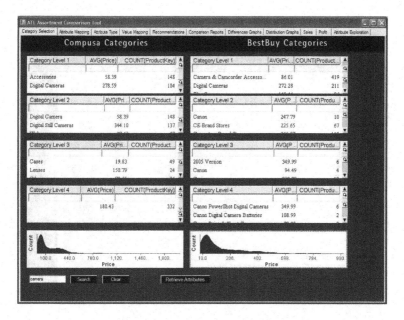

Fig. 1. Mapping of Product Hierarchies: The user selects a set of products considered equivalent when comparing assortments. The bottom histograms show the price distribution in a selection.

egory often is more often guided by practical business constraints than by parsimony. The Assortment Comparison Tool facilitates hierarchy alignment. The user is presented with the hierarchies to be aligned side by side and can select the equivalent categories (see figure 1).

For each assortment, an associated summary histogram shows the number of products and price distribution in the current selection. For example, when comparing digital camera models, the user starts by searching hierarchies for the keyword 'camera'. This search pares down categories at every level of the hierarchy. The user can make other fine tuned selections. For example, after selecting digital cameras and deselecting webcams, the user may inspect the summary graph and find out that there are still some products in the $10-$25 range within those categories. Further inspection reveals that there are some camera accessories subcategories that need to be removed. Thus, the hierarchy alignment process results in the complete alignment of the two hierarchies at the category as well as the subcategory level.

After aligning the categories, the user next considers product attributes. Each product has different types of attributes: some are specific to the product, such as 'megapixels' for digital cameras, and some universal, such as 'price' and 'brand'. As the attributes differ between assortments, they too need to be aligned. While the user can manually align attributes, the tool facilitates the task by suggesting alignments. Identical attributes are trivially matched, but it is less trivial to determine automatically that, for example, 'image formats' and 'file formats' are the same in the context of digital cameras. We use a string matching algorithm to suggest the correct alignments. This algorithm first splits the attributes into tokens (individual words). We use the Levenshtein (edit) distance to

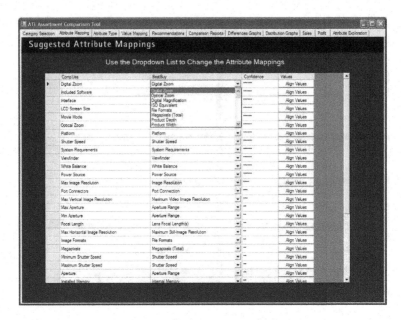

Fig. 2. Mapping Attributes: The user is presented with closest matches and given the opportunity to correct the mappings. The tool displays the confidence score for each match.

compute the similarity of individual tokens. We then employ bipartite graph matching to compute the similarity between two token lists [6]. The best match is presented to the user together with a confidence score. The user can scan the confidence scores to spot aligned attributes in need of manual correction (see figure 2).

While the tool correctly suggests most attribute matches, the user has the opportunity to correct or accept the suggested matches, as well as add matches that were not suggested. This is done by presenting the user with a drop down menu of all attributes that were extracted for each of the assortments.

After the attributes are aligned, they must be classified as categorical type (e.g., 'image formats') or numeric type (e.g., 'megapixels') before their values can be aligned.

Numeric values are aligned automatically and discretized. The Assortment Comparison Tool suggests the type for the attribute, taking into account that numeric attributes may not be immediately identifiable as such. For example, the attribute 'optical zoom' is identified as numerical in spite of the fact that most values do not consist solely of digits (0-9). We consider only values that begin with a digit and strip the units or the suffixes, such as the 'x' in '10x'. We then compute the ratio of consecutive numeric characters to non-numeric characters. Those over a threshold (25%) are considered numeric. If the percentage of numeric values for a given attribute is greater than another threshold (80%), the attribute is labeled numeric, and more aggressive strategies for converting non-numeric values are used. For example, '3.2x (14-45mm lens)' is converted to '3.2'. The values of numeric attributes are optionally discretized. The optical zoom values, for example, are divided into low (2-6), medium (6.5-10), and high (10.7-15). Figure 3 shows the result of numeric value processing.

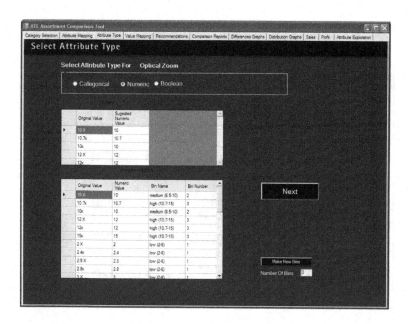

Fig. 3. Attribute Typing and Discretization: Numeric values are cleaned and, optionally, discretized.

Those attributes that do not exceed the above thresholds are automatically considered categorical. Categorical values are aligned in the same way attributes are aligned: the tool presents the user with the best 'guess' alignment based on string matching, but also allows the user to change the mapping.

After the typing of attributes and alignment of values, the assortments as a whole can be compared. The Assortment Comparison Tool makes use of three kinds of assortment data: product data (i.e., attribute-value pairs, the product hierarchy, the number of products in each category, etc.), sales data, and profit data. The tool provides both data visualization capabilities and text reports. While the user can choose to examine the distribution of individual attributes and values in the assortment, the power of the tool comes from aggregating several products by attribute and value and making more high-level statements. All products are divided into high, medium, and low priced bands, and the results are presented overall as well as by individual band. Both difference reports and distribution comparison reports are provided. Difference reports focus on an attribute/value pair in a given price band by comparing the percentages of products in the respective assortment. An automatically generated report may say: 'Compared to competitor X, your store has 16% more high (10-15) optical zoom models.' The same information is also presented graphically (see figure 4).

Distribution reports focus on attributes and the entire distribution for the attribute in a price band. The difference between the distributions is computed using KullbackLeibler divergence [7]. A typical item in a report might be 'Even though the optical zoom distributions for high and medium end products are similar, the low end distribution is different.' The distributions are graphed both for each price band and overall, as can be

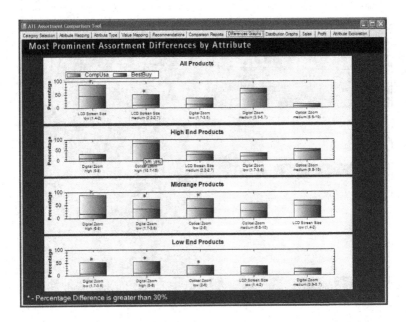

Fig. 4. Differences in assortments by price band. Larger differences are marked with an asterisk.

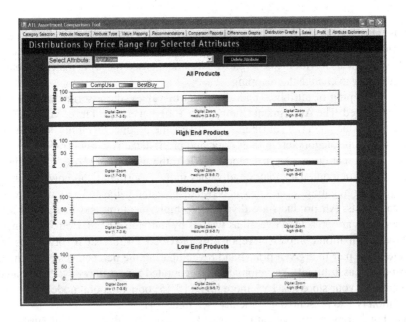

Fig. 5. Attribute/Value distributions by price band: The distributions can be inspected visually instead of relying on KL-divergence

seen in figure 5. By studying both the difference and distribution reports, the user can draw conclusions about how the assortment could be optimized.

The Assortment Comparison Tool itself provides some simple recommendations based on differences in product distribution and profit. For example, if products with certain attributes tend to be high profit items and it turns out that a competitor has more products with the same attributes, it is worth exploring the addition of other products with the same attributes to the assortment. A recommendation might say: 'You might consider increasing the number of products with high (10-15) optical zoom in your assortment. The profits for these products are high, yet you carry 16% fewer models than your competitor'.

In future versions of the tool we intend to provide further forecasting and assortment optimization capabilities, automated support for hierarchy mapping, and construction of a master product hierarchy into which all products would be classified. Such a master hierarchy would allow us, for example, to compare a retailer to *all* its competitors in a comparable way. Another direction of future work is to utilize the user interactions within the context of the tool as training data for mapping hierarchies, attributes and values. As the user corrects the mappings suggested by the tool, we store these mappings and can use them to train mapping algorithms.

5   Conclusions and Future Work

We describe an approach to extract attribute-value pairs from product descriptions and a practical application that is enabled by the extraction algorithm. The extraction system requires very little initial user supervision: using unlabeled data, we automatically extract an initial seed list that serves as training data for the semi-supervised classification algorithm. The extracted attributes and values are linked to form pairs using dependency information and co-location scores. As one example of a practical application of the extracted attribute-value pairs, we described a prototype of an Assortment Comparison Tool that allows retailers to compare their product assortments to those of their competitors. As the comparison is based on attributes and values, we can draw meaningful conclusions at a very fine-grained level.

We plan to focus on adding an interactive step to the extraction algorithm that will allow users to correct extracted pairs as quickly and efficiently as possible. We are experimenting with different active learning algorithms to minimize the number of corrections required to improve the system. We believe that a powerful attribute extraction system can be useful in a wide variety of contexts, as it allows for the normalization of products as attribute-value vectors, which in turn enables granular comparison between products and assortments and improves a variety of applications such as the assortment comparison system described above, but also product recommender systems, price comparison engines, demand forecasting as well as assortment optimization systems.

References

1. Blum, A., Mitchell, T.: Combining labeled and unlabeled data with co-training. In: COLT-98 (1998)
2. Collins, M., Singer, Y.: Unsupervised Models for Named Entity Classification. In: EMNLP/VLC (1999)

3. Ghani, R., Jones, R.: A comparison of efficacy of bootstrapping algorithms for information extraction. In: LREC 2002 Workshop on Linguistic Knowledge Acquisition (2002)
4. Ghani, R., Probst, K., Liu, Y., Krema, M., Fano, A.: Text mining for product attribute extraction. SIGKDD Explorations, Special Issue on Successful Real-World Data Mining Applications
5. Jones, R.: Learning to extract entities from labeled and unlabeled text. Ph.D. Dissertation (2005)
6. Kuhn, H.: The hungarian method for the assignment problem. Naval Research Logistic Quaterly 2, 83–97 (1955)
7. Kullback, S., Leibler, R.: On information and sufficiency. The Annals of Mathematical Statistics 22, 79–86 (1951)
8. Lin, D.: Dependency-based evaluation of MINIPAR. In: Workshop on the Evaluation of Parsing Systems (1998)
9. Liu, B., Hu, M., Cheng, J.: Opinion observer: Analyzing and comparing opinions on the web. In: Proceedings of WWW 2005 (2005)
10. Nigam, K., Ghani, R.: Analyzing the effectiveness and applicability of co-training. In: Proceedings of the Ninth International Conference on Information and Knowledge Management (CIKM-2000) (2000)
11. Peng, F., McCallum, A.: Accurate information extraction from research papers using conditional random fields. In: HLT 2004 (2004)
12. Popescu, A.-M., Etzioni, O.: Extracting product features and opinions from reviews. In: Proceedings of EMNLP 2005 (2005)
13. Seymore, K., McCallum, A., Rosenfeld, R.: Learning hidden markov model structure for information extraction. In: AAI 99 Workshop on Machine Learning for Information Extraction (1999)

Discovering User Profiles from Semantically Indexed Scientific Papers

Giovanni Semeraro, Pierpaolo Basile, Marco de Gemmis, and Pasquale Lops

Department of Informatics
University of Bari
Via E. Orabona, 4 - 70125 Bari, Italia
{semeraro,basilepp,degemmis,lops}@di.uniba.it

Abstract. Typically, personalized information recommendation services automatically infer the user profile, a structured model of the user interests, from documents that were already deemed relevant by the user. We present an approach based on Word Sense Disambiguation (WSD) for the extraction of user profiles from documents. This approach relies on a knowledge-based WSD algorithm, called JIGSAW, for the semantic indexing of documents: JIGSAW exploits the WordNet lexical database to select, among all the possible meanings (*senses*) of a polysemous word, the correct one. Semantically indexed documents are used to train a naïve Bayes learner that infers "semantic", *sense-based* user profiles as binary text classifiers (user-likes and user-dislikes).

Two empirical evaluations are described in the paper. In the first experimental session, JIGSAW has been evaluated according to the parameters of the SENSEVAL-3 initiative, that provides a forum where the WSD systems are assessed against disambiguated datasets. The goal of the second empirical evaluation has been to measure the accuracy of the user profiles in selecting relevant documents to be recommended. Performance of classical keyword-based profiles has been compared to that of sense-based profiles in the task of recommending scientific papers. The results show that sense-based profiles outperform keyword-based ones.

Keywords: User Profiling, Text Categorization, Word Sense Disambiguation, WordNet, Text Mining for Information Retrieval.

1 Introduction

The amount of information available on the Web and in Digital Libraries is increasing over time. The role of user modeling and personalized information access is increasing: Users need a personalized support in sifting through large amounts of retrieved information according to their interests. Information filtering and retrieval systems relying on this idea adapt their behavior to individual users by learning their preferences during the interaction in order to construct a *user profile* that can be exploited later in the search process. Traditional keyword-based approaches are unable to capture the *semantics* of the

B. Berendt et al. (Eds.): WebMine 2006, LNAI 4737, pp. 61–81, 2007.

user interests. They are primarily driven by a string-matching operation: If a string is found in both the profile and the document, a match is made and the document is considered as relevant. String matching suffers from problems of *polysemy*, the presence of multiple meanings for one word, and *synonymy*, multiple words having the same meaning. The result is that, due to synonymy, relevant information can be missed if the profile does not contain the exact keywords in the documents while, due to polysemy, wrong documents could be deemed as relevant. These problems call for alternative methods able to learn more accurate profiles that capture concepts expressing users' interests from relevant documents. These *semantic* profiles will contain references to concepts defined in lexicons or ontologies. This paper describes an approach in which semantic user profiles are obtained by machine learning techniques integrated with a word sense disambiguation (WSD) strategy based on the WordNet lexical database [17, 6]. We consider the problem of learning user profiles as a binary text categorization task: Each document has to be classified as interesting or not compared to the user preferences. Therefore, the set of categories is $C = \{c_+, c_-\}$, where c_+ is the positive class, *user-likes*, and c_- the negative one, *user-dislikes*. There are several ways in which documents can be represented in order to be used as a basis for the learning component and there exists a variety of machine learning methods that could be used for inferring user profiles. The proposed strategy consists of two steps. In the first one, the JIGSAW WSD algorithm is used to assign the most appropriate meaning to each word in the documents to be indexed by exploiting the lexical database WordNet as a sense repository. In the second step, a naïve Bayes approach learns *sense-based* user profiles as binary text classifiers from disambiguated documents. The paper is organized as follows: After a brief discussion about the main works related to our research, we give in Section 3 some details about WordNet, before describing the JIGSAW algorithm in Section 4. Section 5 presents the formal model we adopted for the semantic indexing of documents by using WordNet senses. Section 6 describes the naïve bayes text categorization method we adopted to build *WordNet-based* user profiles. The method is implemented by our content-based profiling system ITem Recommender (ITR). An experimental evaluation has been carried out to evaluate our approach by comparing the performance of keyword-based profiles with that of sense-based profiles. The main results are presented in Section 7. Conclusions and future work are discussed in the last Section.

2   Related Work

Our research was mainly inspired by the following works. *LIBRA* [18] adopts a Bayesian classifier to produce content-based book recommendations by exploiting product descriptions obtained from Web pages. The main limitation of this work is that keywords are used to represent documents.

SiteIF [10] exploits a *sense-based* representation to build a user profile as a semantic network whose nodes represent senses of the words in documents requested by the user. In the modeling phase, SiteIF considers the synsets (senses in WordNet) in the documents browsed during a user navigation session. Synsets are recognized by Word Domain Disambiguation (WDD), which is a variant of WSD where, for each word in a text, a domain label (Literature, Religion, . . .) is chosen instead of a sense label. In *SiteIF*, for each *noun*, the synsets associated to the proposed domain are selected and added to the document representation. The system builds the semantic net by including the synsets occurring in the browsed collection in the nodes of the net and by assigning each node with a score that is inversely proportional to its frequency over all the corpus. Arcs between nodes represent the co-occurrence of two synsets in a document. Our approach is different both in the disambiguation process and in the construction of the user profile. We do not perform WDD on nouns, but we try to assign the most appropriate synset to each word in a document. As regards the user model, we learn a probability distribution of the senses found in the corpus of the documents rated by the user.

A different approach to identify concepts from existing terms in documents is Latent Semantic Indexing (LSI) [4], which does not rely on any knowledge base. This technique compresses document vectors into vectors of a lower-dimensional space whose dimensions are obtained as combinations of the original dimensions by looking at their patterns of co-occurrence. One characteristic of LSI is that the newly obtained dimensions are not intuitively interpretable. In our approach, we try to identify *definite* WordNet concepts in order to include them in the index.

OntoSeek [7] is a knowledge-retrieval system for online yellow pages and product catalogs which explored the role of linguistic ontologies in retrieval systems. The approach has shown that structured content representations coupled with linguistic ontologies can increase both recall and precision. By taking into account the lessons learned by the previously cited works, we conceived our ITR system as a text classifier able to learn a bayesian profile from documents subdivided into slots and indexed by using *senses*, instead of keywords.

The strategy we propose in order to shift from a keyword-based document representation to a sense-based one representation is *to integrate lexical knowledge in the indexing step of training documents.* Several methods have been proposed to accomplish this task. Scott and Matwin [21] proposed to expand each word in the training set with *all* the synonyms for it in WordNet in order to avoid WSD. This approach has shown a decrease of effectiveness in the obtained classifier, mostly due to the word ambiguity problem. More recent work [12, 2] provided a sound experimental evidence of the usefulness of embedding WSD in classification tasks, especially when a limited number of labeled examples is given, as in user profiling tasks. In [12], WordNet is used as a hierarchical thesaurus both for WSD and for classification, while our approach relies on the hypernymy/hyponymy relation only for the computation of the semantic similarity between synsets. In [2] the authors experiment with various settings for mapping words to senses: No WSD, most frequent sense as provided by WordNet, WSD based on context. They found positive results on the Reuters 25178[1], the OHSUMED[2] and the FAODOC[3] corpus. The improved results can be ascribed to multi-word expression detection, synonym conflation, and to the exploitation of ontology structures for generalization. In our knowledge-based WSD approach, generalization is used only to detect the most specific subsumer of two concepts for computing semantic similarity.

The most successful approaches for *all words* WSD rely on information drawn from annotated corpora. The system by Decadt [3] uses two cascaded memory-based classifiers, combined with the use of a genetic algorithm for joint parameter optimization and feature selection. A separate "word expert" is learned for each ambiguous word, using a concatenated corpus of English sense tagged texts, including SemCor[4], SENSEVAL datasets, and a corpus built from Word-Net examples. The performance of this system on the SENSEVAL-3 English all words dataset was evaluated at 65.2%. The system developed by Yuret [26] is based on statistical models built on SemCor and WordNet, for an overall disambiguation accuracy of 64.1%. Both previously cited systems use supervised learning methods, where each sense-tagged occurrence of a particular word is transformed into a feature vector, which is exploited by an automated learning process. The main limitation of these supervised algorithms is that they need a tagged corpus for training data. The applicability of these approaches is limited only to those words for which sense tagged data is available, and their accuracy is strongly connected to the amount of labeled data available at hand.

In an attempt to overcome these limitations, a minimally supervised approach is adopted by the SENSELEARNER system [16], which learns general semantic models for various word categories, starting with a relatively small sense-annotated corpus. The results obtained by SENSELEARNER on both SENSEVAL-2 and SENSEVAL-3 data sets were proved competitive with the best published results on the same data sets.

On the other hand, knowledge-based approaches reach lower accuracy levels than supervised methods (even if results are not directly comparable because supervised approaches take advantages of large training data sets), but they do not need any training data. Furthermore, the increasing availability of large-scale, rich lexical knowledge resources seems to provide new challenges to knowledge-based approaches. Two interesting works in this direction are proposed by Navigli and Velardi [19] and Mihalcea [14].

In this paper we propose a detailed description of a knowledge-based algorithm and provide experimental results which are taken as a starting point for future improvements.

---

[1] http://about.reuters.com/researchandstandards/corpus/
[2] http://www.ltg.ed.ac.uk/disp/resources/
[3] http://www4.fao.org/faobib/index.html
[4] http://www.cs.unt.edu/~rada/downloads.html#semcor

3   Using WordNet for Word Sense Disambiguation and Semantic Indexing

Textual documents cannot be directly interpreted by machine learning algorithms. An indexing procedure that maps a document d_i into a compact representation of its content must be applied. In the classical *bag-of-words* (BOW) approach, each document is represented as a feature vector counting the number of occurrences of different words as features [22]. We extend the BOW model to a model in which each document is represented by the senses conveyed by the words in its content, together with their respective occurrences. Here, "sense" is used as a synonym of "meaning". This semantic indexing model is exploited by the machine learning algorithm to build semantic user profiles (Section 6). Any implementation of a sense-based document indexing must solve the problem that, while words occur in a document, meanings do not, since they are often hidden in the context. As a consequence, a procedure is needed for assigning senses to words. This task, known as *word sense disambiguation* (WSD), consists in determining which of the senses of an ambiguous word is invoked in a particular use of that word [11]. Therefore, the goal of a WSD algorithm is to associate each word w_i occurring in a document d with its appropriate meaning or sense s, by exploiting the *context* C in which w_i is found, commonly defined as a set of words that precede and follow w_i. The sense s is selected from a predefined set of possibilities, usually known as *sense inventory*. In the proposed algorithm, WordNet is used as a the sense repository.

WordNet is a semantic lexicon for the English language. It groups English words into sets of synonyms called *synsets*, provides short general definitions, and records the various semantic relations between these synonym sets. The purpose is twofold: To produce a combination of dictionary and thesaurus that is more intuitively usable, and to support automatic text analysis and artificial intelligence applications. WordNet distinguishes between nouns, verbs, adjectives and adverbs because they follow different grammatical rules. Every synset contains a group of synonymous words or collocations; different senses of a word are in different synsets. The meaning of the synsets is further clarified with short defining glosses. A typical example synset with gloss is:

good, right, ripe – (most suitable or right for a particular purpose; "a good time to plant tomatoes"; "the right time to act"; "the time is ripe for great sociological changes")

In our algorithm, we use the hypernymy/hyponymy semantic relation for nouns and verbs. WordNet also provides the polysemy count of a word as the number of synsets that contain that word. If a word participates in several synsets, then typically some senses are more common than others. WordNet quantifies this by the frequency score. Several sample texts have all words semantically tagged with the corresponding synset, and then a count indicating how often a word occurs in a specific sense is provided.

4   The JIGSAW Algorithm for Word Sense Disambiguation

Since the performance of the WSD algorithms change in accordance to the part-of-speech (POS) of the word to be disambiguated, the proposed JIGSAW algorithm combines three different strategies to disambiguate nouns, verbs, adjectives and adverbs. An adaptation of the Lesk dictionary-based WSD algorithm has been used to disambiguate adjectives and adverbs [1], an adaptation of the Resnik algorithm has been used to disambiguate nouns [20], while the algorithm we developed for disambiguating verbs exploits the nouns in the context of the verb and the nouns both in the glosses and in the phrases that WordNet utilizes to describe the usage of the verb. The algorithm disambiguates only words which belong to at least one synset. JIGSAW takes as input a document $d = (w_1, w_2, \ldots, w_h)$ encoded as a list of words in order of their appearance, and will output a list of WordNet synsets $X = (s_1, s_2, \ldots, s_k)$, $(k \leq h)$, in which each element s_j is obtained by disambiguating the *target word* w_i based on the information obtained from WordNet about a few immediately surrounding words. Notice that $k \leq h$ because either some words could not be found in WordNet, such as proper names, or because of bigram recognition. We define the *context* C of the target word to be a window of n words to the left and another n words to the right, for a total of $2n$ surrounding words. If w_i is near the beginning or the end of d, we consider additional words from the other direction. The algorithm is based on three different procedures for nouns, verbs, adverbs and adjectives, called $JIGSAW_{nouns}$, $JIGSAW_{verbs}$, $JIGSAW_{others}$ respectively. The POS tag of each word is computed by the HMM-based tagger ACOPOST t3[5]. More details for each one of the above mentioned procedures follow.

4.1 $JIGSAW_{nouns}$

The procedure is obtained by making some variations to the algorithm designed by Resnik [20] for disambiguating noun groups. Given a set of nouns $W = \{w_1, w_2, \ldots, w_n\}$, obtained from document d, with each w_i having an associated sense inventory $S_i = \{s_{i1}, s_{i2}, \ldots, s_{ik}\}$ of possible senses, the goal is to associate each w_i with the most appropriate sense $s_{ih} \in S_i$, according to the *similarity* of w_i with the other words in W (the context for w_i). The idea is to define a function $\varphi(w_i, s_{ij})$, $w_i \in W$, $s_{ij} \in S_i$ that computes a value in $[0, 1]$, representing the confidence with which sense s_{ij} can be associated with w_i.

The intuition behind this algorithm is essentially the same intuition exploited by Lesk [9] and others: The most plausible assignment of senses to multiple co-occurring words is the one that maximizes *relatedness* of meaning among the chosen senses. $JIGSAW_{nouns}$ differs from the original algorithm by Resnik [20] in the similarity measure used to compute the relatedness of two senses. We adopted the Leacock-Chodorow measure [8], which is based on the length of the path between concepts in an IS-A hierarchy. The similarity between two

---

[5] http://acopost.sourceforge.net/

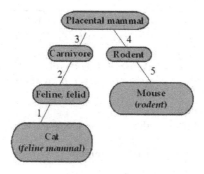

Fig. 1. A fragment of the WordNet hierarchy

synsets, s_1 and s_2, is inversely proportional to their distance in the WordNet IS-A hierarchy. The distance is computed by:

- finding the *most specific subsumer* (MSS) of s_1 and s_2 (each ancestor of both s_1 and s_2 in the WordNet hierarchy is a subsumer, the MSS is the one at the lowest level);
- counting the number of nodes in the path between s_1 and s_2 that traverses their MSS.

For example, Figure 1 shows that the length of the path between *cat* (feline mammal) and *mouse* (rodent) by traversing *placental mammal* is 5. The similarity between two *synsets* is computed by the function described in Algorithm 1. In the example, SYNSIM(*feline mammal, rodent*) = 0.806. We extended this measure by introducing a parameter k that restricts the search for the MSS to k ancestors (i.e. that climbs the WordNet IS-A hierarchy until either it finds the MSS or $k + 1$ ancestors of both s_1 and s_2 have been explored). This guarantees that "too abstract" (i.e. "less informative") MSSs will be ignored.

In the example, SYNSIM(*feline mammal, rodent, 4*) = 0.806, SYNSIM(*feline mammal, rodent, 2*) = 0. Before describing the whole $JIGSAW_{nouns}$ procedure, we need to define the semantic similarity between two *words* w_1 and w_2, as in Algorithm 3.

Algorithm 1. The Leacock-Chodorow similarity function between synsets

1: **function** SYNSIM(s_1, s_2)   ▷ The similarity for the synsets s_1 and s_2
2:     $MSS \leftarrow MSS(s_1, s_2)$ ▷ MSS(s_1,s_2) returns the most specific subsumer
            between s_1 and s_2
3:     $N_p \leftarrow$ the number of nodes in the shortest path p from s_1 to s_2 by traversing
            MSS
4:     $D \leftarrow$ maximum depth of the taxonomy
5:     $r \leftarrow -log(N_p/2D)$
6:     **return** r
7: **end function**

Algorithm 2. The modified similarity function between synsets

---

function SYNSIM(s_1, s_2, k)   ▷ The similarity for the synsets s_1 and s_2, search for MSS restricted to k ancestors

$MSS \leftarrow MSS(s_1, s_2, k)$ ▷ MSS(s_1,s_2,k) returns the most specific subsumer between s_1 and s_2, k ancestors explored, at the most

$r \leftarrow 0$

if MSS not null **then**

N_p ←the number of nodes in the shortest path p from a to b by traversing MSS

D ←maximum depth of the taxonomy

$r \leftarrow -log(N_p/2D)$

end if

return r

end function

Algorithm 3. The similarity function between words

---

function SIM(w_1, w_2, k)   ▷ The similarity for the words w_1 and w_2, search for MSS restricted to k ancestors

$S_1 \leftarrow \{s_{11}, s_{12}, \ldots, s_{1n}\}$ ▷ sense inventory for w_1

$S_2 \leftarrow \{s_{21}, s_{22}, \ldots, s_{2m}\}$ ▷ sense inventory for w_2

for all $s_{1i} \in S_1$ and $s_{2j} \in S_2$ **do**

$MSS_{ij} \leftarrow MSS(s_{1i}, s_{2j}, k)$ ▷ returns the most specific subsumer between s_{1i} and s_{2j}, k ancestors explored, at the most

N_{ij} ←the number of nodes in the shortest path from s_{1i} to s_{2j} by traversing MSS_{ij}

end for

$MSS_{xy} \leftarrow$ the MSS that minimizes N_{ij} ▷ MSS_{xy} is the synset that subsumes both w_1 and w_2, in any sense of either word

$r \leftarrow$ SYNSIM(s_{1x}, s_{2y}, k)   ▷ the maximum value of similarity for all possible pairs of senses associated with w_1 and w_2

return r

end function

In addition to the semantic similarity function, the $JIGSAW_{nouns}$ differs from the Resnik algorithm in the use of:

- a Gaussian factor which takes into account the distance between the words in the text to be disambiguated. This factor is used in the algorithm to amplify (or reduce) the similarity score computed between words: The closer words are to each other, the more likely they are to be relevant to the disambiguation of each other (line 3 of Algorithm 4).
- a factor which gives more importance to the synsets that are more commonly used than others, according to the frequency score (Section 3);
- a *parametrized* search for the MSS between two concepts. In this way, the search is limited to a certain number of ancestors and consequently more abstract subsumers are excluded from the computation.

Algorithm 4 describes the complete procedure for the disambiguation of nouns.

Algorithm 4. The procedure for disambiguating nouns derived from the algorithm by Resnik

---

1: **procedure** $JIGSAW_{nouns}(W, depth1, depth2)$ ▷ finds the proper synset for each polysemous noun in the set $W = \{w_1, w_2, \ldots, w_n\}$, $depth1$ and $depth2$ are used in the computation of MSS

2:      **for all** $w_i, w_j \in W$ **do**

3:         $sim \leftarrow \text{SIM}(w_i, w_j, depth1) * Gauss(position(w_i), position(w_j))$ ▷ $Gauss(x,y)$ is a Gaussian function which takes into account the difference between the positions of w_i and w_j

4:         $MSS_{ij} \leftarrow MSS(w_i, w_j, depth2)$ ▷ MSS_{ij} is the most specific subsumer between w_i and w_j, search for MSS restricted to $depth2$ ancestors

5:         **for all** $s_{ik} \in S_i$ **do**

6:            **if** is-ancestor(MSS_{ij}, s_{ik}) **then**   ▷ if MSS_{ij} is an ancestor of s_{ik}

7:               $support_{ik} \leftarrow support_{ik} + sim$

8:            **end if**

9:         **end for**

10:        **for all** $s_{jh} \in S_j$ **do**

11:           **if** is-ancestor(MSS_{ij}, s_{jh}) **then**

12:             $support_{jh} \leftarrow support_{jh} + sim$

13:           **end if**

14:        **end for**

15:        $normalization_i \leftarrow normalization_i + sim$

16:        $normalization_j \leftarrow normalization_j + sim$

17:      **end for**

18:      **for all** $w_i \in W$ **do**

19:        **for all** $s_{ik} \in S_i$ **do**

20:           **if** $normalization_i > 0$ **then**

21:             $\varphi(i, k) \leftarrow \alpha * support_{ik}/normalization_i + \beta * R(k)$

22:           **else**

23:             $\varphi(i, k) \leftarrow \alpha/|S_i| + \beta * R(k)$

24:           **end if**

25:        **end for**

26:      **end for**

27: **end procedure**

This algorithm considers the words in W pairwise. For each pair (w_i, w_j), the most specific subsumer MSS_{ij} is identified, by reducing the search to $depth1$ ancestors, at the most. Then, the similarity $\text{SIM}(w_i, w_j, depth2)$ between the two words is computed, by reducing the search for the MSS to $depth2$ ancestors, at the most. MSS_{ij} is considered *as supporting evidence* for those synsets s_{ik} in S_i and s_{jh} in S_j that are descendants of MSS_{ij}. The amount of support contributed by the pairwise comparison is the similarity value computed according to the function described in Algorithm 3, weighted by a gaussian factor that takes into account the position of w_i and w_j in W (the shorter is the distance between the words, the higher is the weight). The value $\varphi(i, k)$ assigned to each candidate synset s_{ik} for the word w_i is the sum of two elements. The first one is the proportion of support it received, out of the support possible, computed as: $support_{ik}/normalization_i$ in the pseudocode. The other element that contributes to $\varphi(i, k)$ is a factor $R(k)$ that takes into account rank of s_{ik} in WordNet, i. e. how common sense s_{ik} is for the word w_i. $R(k)$ is computed as:

$$R(k) = 1 - 0.8 * \frac{k}{n-1} \qquad (1)$$

where n is the cardinality of the sense inventory S_i for w_i, and k is the rank of s_{ik} in S_i, starting from 0. Both elements are weighted by two parameters: α, which controls the contribution given to $\varphi(i, k)$ by the normalized support, and β, which controls the contribution given by the rank of s_{ik}. The value for $\varphi(i, k)$ is computed as in Algorithm 4 (lines 20-23). We set $\alpha = 0.7$ and $\beta = 0.3$. The synset assigned to each word in W is the one with the highest φ value. Notice that we used two different parameters, $depth1$ and $depth2$ for setting the maximum depth for the search of the MSS: $depth1$ limits the search for the MSS computed in the similarity function, while $depth2$ limits the computation of the MSS used for assigning support to candidate synsets. For example, by setting $depth1 = 6$ and $depth2 = 3$, we allow the algorithm to ascend the WordNet hierarchy for searching the MSS until a high level of abstraction, but we impose a stronger requirement for the computation of the MSS used for assigning support. This means that only synsets which are descendants of very "specific" MSS will receive support.

4.2 $JISAW_{verbs}$

Before describing the $JIGSAW_{verbs}$ procedure, the *description* of a synset must be defined. It is the string obtained by concatenating the gloss and the sentences that WordNet uses to explain the usage of a word.

For example, the gloss for the synset corresponding to the sense n.2 of the verb *look* ($\{look, appear, seem\}$) is "*give a certain impression or have a certain outward aspect*", while some examples of usage of the verb are: "*She seems to be sleeping*"; "*This appears to be a very difficult problem*". The description of the synset is "*give a certain impression or have a certain outward aspect She seems to be sleeping This appears to be a very difficult problem*". First, the $JIGSAW_{verbs}$ includes in the context C for the target verb w_i all the nouns in the window of $2n$ words surrounding w_i. For each candidate synset s_{ik} of w_i, the algorithm computes $nouns(i, k)$, that is the set of nouns in the description for s_{ik}. In the above example, $nouns(look, 2) = \{impression, aspect, problem\}$. Then, for each w_j in C and each synset s_{ik}, the following value is computed:

$$max_{jk} = max_{w_l \in nouns(i,k)} \{\text{SIM}(w_j, w_l, depth)\} \qquad (2)$$

where $\text{SIM}(w_j, w_l, depth)$ is defined as in Algorithm 3. In other words, max_{jk} is the highest similarity value for w_j, wrt the nouns related to the k-th sense for w_i. Finally, an overall similarity score among s_{ik} and the whole context C is computed:

$$\varphi(i,k) = R(k) \cdot \frac{\sum_{w_j \in C} Gauss(position(w_i), position(w_j)) \cdot max_{jk}}{\sum_h Gauss(position(w_i), position(w_h))} \quad (3)$$

where $R(k)$ is defined as in Equation 1 and $Gauss(position(w_i), position(w_j))$ is the same Gaussian factor used in $JIGSAW nouns$ that gives a higher weight to words closer to the target word. The synset assigned to w_i is the one with the highest φ value. Algorithm 5 provides a detailed description of the procedure.

4.3 $JIGSAW_{others}$

This procedure is based on the WSD algorithm proposed in [1]. The idea is to compare the glosses of each candidate sense for the target word to the glosses of all the words in its context. Let W_i be the sense inventory for the target word w_i. For each $s_{ik} \in W_i$, $JIGSAW_{others}$ computes the string $targetGloss_{ik}$ that contains the words in the gloss of s_{ik}. Then, the procedure computes the string $contextGloss_i$, which contains the words in the glosses of all the synsets corresponding to each word in the context for w_i. Finally, the procedure computes the overlap between $contextGloss_i$ and $targetGloss_{ik}$, and assigns the synset with the highest overlap score to w_i. This score is computed by counting the words that occur both in $targetGloss_{ik}$ and in $contextGloss_i$.

4.4 Experiments

JIGSAW was evaluated according to the parameters of the SENSEVAL-3 initiative, that provides a forum where WSD systems are assessed against disambiguated datasets. The "All Words Task" for English and the "English Sample Task" were chosen.

Typical measures adopted to evaluate WSD algorithms are: *Precision*, defined as the proportion of disambiguated words that were correctly disambiguated, and *Recall*, which is the proportion of words disambiguated correctly. JIGSAW renounces to disambiguate the target word only whether it is not included in WordNet, but since this case never happened in the datasets used in the evaluation, all the target words were disambiguated. As a consequence, JIGSAW precision and recall coincide (we report only precision in the tables).

English All Words Task. This task evaluates the ability of the WSD algorithm to disambiguate all the words in a text. The dataset consists in approximately 5000 words of coherent Penn Treebank text with WordNet 1.7.1 tags.

The experiment was designed by using the following parameters:

- MSS-depth: maximum depth for searching the MSS used for assigning support. If the MSS of two synsets or words was not found at the fixed MSS-depth, the MSS is *null*, thus any support is assigned;
- SIM-depth: maximum depth for searching the MSS in the computation of the similarity score. If the MSS of two synset was not found at the fixed SIM-depth, the similarity is set to 0;

Algorithm 5. The procedure for the disambiguation of verbs

1: **procedure** $JIGSAW_{verbs}(w_i, d, depth)$ ▷ finds the proper synset of a polysemous
    verb w_i in document d
2:    $C \leftarrow \{w_1, ..., w_n\}$ ▷ C is the context for w_i. For example,
    $C = \{w_1, w_2, w_4, w_5\}$, if the sequence of words $\{w_1, w_2, w_3, w_4, w_5\}$ occurs
    in d, w_3 being the target verb, w_j being nouns, $j \neq 3$
3:    $S_i \leftarrow \{s_{i1}, ...s_{im}\}$ ▷ S_i is sense inventory for w_i, that is the set of all candidate
    synsets for w_i returned by WordNet
4:    $s \leftarrow null$ ▷ s is the synset to be returned
5:    $score \leftarrow -MAXDOUBLE$ ▷ $score$ is the similarity score assigned to s
6:    $p \leftarrow 1$ ▷ pos is the position of the synsets in W_i
7:    **for all** $s_{ik} \in S_i$ **do**
8:        $max \leftarrow \{max_{1k}, ..., max_{nk}\}$
9:        $nouns(i, k) \leftarrow \{noun_1, ..., noun_z\}$ ▷ $nouns(i, k)$ is the set of all nouns in
        the description of s_{ik}
10:      $sumGauss \leftarrow 0$
11:      $sumTot \leftarrow 0$
12:      **for all** $w_j \in C$ **do**   ▷ computation of the similarity between C and s_{ik}
13:        $max_{jk} \leftarrow 0$ ▷ max_{jk} is the highest similarity value for w_j, wrt the
        nouns related to the k-th sense for w_i.
14:        $sumGauss \leftarrow$Gauss(position(w_i),position(w_j))   ▷
        Gaussian function which takes into account the difference between
        the positions of the nouns in d
15:        **for all** $noun_l \in nouns(i, k)$ **do**
16:          $sim \leftarrow$ SYNSIM(w_j,$noun_l$,depth)   ▷ sim is the similarity between
          the j-th noun in C and l-th noun in nouns(i,k)
17:          **if** $sim > max_{jk}$ **then**
18:            $max_{jk} \leftarrow sim$
19:          **end if**
20:        **end for**
21:      **end for**
22:      **for all** $w_j \in C$ **do**
23:        $sumTot \leftarrow$sumTot + Gauss(position(w_i),position(w_j))*max_{jk}
24:      **end for**
25:      $sumTot \leftarrow$sumTot/$sumGauss$
26:      $\varphi(i, k) \leftarrow R(k)$*sumTot   ▷ R(k) is defined as in $JIGSAW_{nouns}$
27:      **if** $\varphi(i, k) > score$ **then**
28:        $score \leftarrow \varphi(i, k)$
29:        $p \leftarrow k$
30:      **end if**
31:    **end for**
32:    $s \leftarrow s_{ip}$
33:    **return** s
34: **end procedure**

In this experiment, the context size was set to 18 because the goal was to stress
the algorithm by exploiting a large context. In Table 1 we report the results of
the experiment.

Table 1. English All Words Task results. Last column reports the time required by the computation.

MSS-depth	SIM-depth	Precision	Time
2	3	0.494	12m
2	6	0.496	15m
2	MAX	0.520	18m
3	3	0.494	11m
3	6	0.486	14m
3	MAX	0.486	18m
4	3	0.481	11m
4	6	0.483	15m
4	MAX	0.484	16m
MAX	3	0.466	18m
MAX	6	0.467	18m
MAX	MAX	0.467	17m

It can be noticed that 52% of Precision is obtained when MSS-depth= 2 and SIM-depth= MAX (the maximum depth of the WordNet taxonomy). We can attribute this result to the strict requirement set by $MSS - depth$: More abstract subsumers are excluded from the computation, because they are "less informative". On the other hand, when $SIM - depth$ is set to a high level of abstraction, the similarity scores can contribute to the computation of support, even if a high level of generalization is reached in the taxonomy.

English Sample Task. This task evaluates the ability of the WSD algorithm to disambiguate a single word in a particular context. In spite of this task is typically used to evaluate supervised algorithms, we want to measure the performance of our algorithm in this more difficult task (for knowledge-based approaches).

The data have been collected via the Open Mind Word Expert (OMWE) interface[15]. When Senseval-3 took place, the dataset had enough data for about 60 ambiguous nouns, adjectives, and verbs. The dataset uses WordNet 1.7.1 as sense repository for nouns and adjectives, and Wordsmyth[6] for verbs.

Table 2 and Table 3 report the results for verbs and nouns respectively. A remarkable observation is that, even if we increase SIM-depth, any significant improvement in precision is obtained for both nouns and verbs. On the other hand, $JIGSAW_{nouns}$ precision improves from 0.260 to 0.319 when a stronger requirement is set on MSS-depth. It's a reasonable result, because a high MSS-depth value can introduce a high-level concept which is hypernym of a lot of concepts. This effect produces a worse estimation of the MSS.

The results show that JIGSAW performs comparably to other knowledge-based algorithms in SENSEVAL-3 competition.

---

[6] http://www.wordsmyth.net/

Table 2. English Sample Task results for $JIGSAW_{verbs}$

SIM-depth	Context Size	Precision	Time
3	6	0.398	12m
3	12	0.405	12m
3	18	0.410	12m
6	6	0.400	11m
6	12	0.404	12m
6	18	0.410	16m
MAX	6	0.399	10m
MAX	12	0.404	15m
MAX	18	0.410	18m

Table 3. English Sample Task results for $JIGSAW_{nouns}$

Context size	SIM-depth	MSS-depth	Precision	Time
12	3	2	0.315	11m
18	3	2	0.319	11m
12	6	2	0.315	10m
18	6	2	0.319	12m
12	MAX	2	0.315	12m
18	MAX	2	0.319	11m
12	3	3	0.299	11m
18	3	3	0.298	10m
12	6	3	0.299	11m
18	6	3	0.299	13m
12	MAX	3	0.299	11m
18	MAX	3	0.299	13m
12	3	4	0.257	11m
18	3	4	0.261	12m
12	6	4	0.257	12m
18	6	4	0.260	13m
12	MAX	4	0.260	12m
18	MAX	4	0.260	12m

5 Keyword-Based and Synset-Based Document Representation

The WSD procedure is fundamental to obtain a synset-based vector space representation that we called Bag-Of-Synsets (BOS). In this model, a vector of synsets corresponds to a document, instead of a vector of keywords. In our approach, each document is structured into *slots*, each slot being a textual field corresponding to a specific feature of the document. For example, in our experiments, documents are scientific papers represented by three slots: *Title*, *Authors* (only names), and *Abstract*. The text in each slot is represented according to the BOS model by counting separately the occurrences of a synset in the slots in which it occurs. Assume that we have a collection of N documents. Let m be the index of the slot, for $n = 1, 2, ..., N$, the n-th document is reduced to three bags of synsets, one for each slot:

$$d_n^m = \langle t_{n1}^m, t_{n2}^m, \ldots, t_{nD_{nm}}^m \rangle$$

where t_{nk}^m is the k-th synset in slot s_m of document d_n and D_{nm} is the total number of synsets appearing in the m-th slot of document d_n. For all n, k and m, $t_{nk}^m \in V_m$, which is the vocabulary for the slot s_m (the set of all different synsets found in slot s_m). Document d_n is finally represented in the vector space by three synset-frequency vectors:

$$f_n^m = \langle w_{n1}^m, w_{n2}^m, \ldots, w_{nD_{nm}}^m \rangle$$

where w_{nk}^m is the weight of the synset t_k in the slot s_m of document d_n and can be computed in different ways: It can be simply the number of times synset t_k occurs in slot s_m, as we used in our experiments, or a more complex TF-IDF score. Our hypothesis is that the proposed document representation helps to obtain profiles able to recommend documents semantically closer to the user interests. The difference wrt keyword-based profiles is that synset unique identifiers are used instead of words. In the next section, we describe the learning algorithm adopted to build semantic user profiles, starting from the BOS document representation.

6   A Naïve Bayes Method for User Profiling

Naïve Bayes is a probabilistic approach to inductive learning. The learned probabilistic model estimates the *a posteriori* probability, $P(c_j|d_i)$, of document d_i belonging to class c_j. To classify a document d_i, the class with the highest probability is selected. As a working model for the naïve Bayes classifier, we use the multinomial event model [13]:

$$P(c_j|d_i) = P(c_j) \prod_{w \in V_{d_i}} P(t_k|c_j)^{N(d_i, t_k)} \qquad (4)$$

where $N(d_i, t_k)$ is defined as the number of times word or token t_k appeared in document d_i. Notice that rather than getting the product of all distinct words in the corpus, V, we only use the subset of the vocabulary, V_{d_i}, containing the words that appear in the document d_i.

Since each instance is encoded as a vector of BOS, one for each slot, Equation (4) becomes:

$$P(c_j|d_i) = \frac{P(c_j)}{P(d_i)} \prod_{m=1}^{|S|} \prod_{k=1}^{|b_{im}|} P(t_k|c_j, s_m)^{n_{kim}} \qquad (5)$$

where $S = \{s_1, s_2, \ldots, s_{|S|}\}$ is the set of slots, b_{im} is the BOS in the slot s_m of the instance d_i, n_{kim} is the number of occurrences of the synset t_k in b_{im}. Our ITR profiling system implements this approach to classify documents as interesting or uninteresting for a particular user. To calculate (5), we only need to estimate $P(c_j)$ and $P(t_k|c_j, s_m)$ in the training phase of the system. The documents used to train the system belong to a collection of scientific papers accepted to the 2002-2004 editions of the International Semantic Web Conference (ISWC). Ratings on these documents, obtained from real users, were recorded on a discrete scale from 1 to 5 (see Section 7 for a detailed description of the dataset). An instance labeled with a rating r, $r = 1$ or $r = 2$ belongs to class c_- (user-dislikes); if $r = 4$ or $r = 5$ then the instance belongs to class c_+ (user-likes); rating $r = 3$ is neutral. Each rating was normalized to obtain values ranging between 0 and 1:

$$w_+^i = \frac{r-1}{MAX-1}; \qquad w_-^i = 1 - w_+^i \qquad (6)$$

where MAX is the maximum rating that can be assigned to an instance. The weights in (6) are used for weighting the occurrences of a synset in a document and to estimate the probability terms from the training set TR. The prior probabilities of the classes are computed according to the following equation:

$$\hat{P}(c_j) = \frac{\sum_{i=1}^{|TR|} w_j^i + 1}{|TR| + 2} \qquad (7)$$

Witten-Bell smoothing [24] has been adopted to compute $P(t_k|c_j, s_m)$, by taking into account that documents are structured into slots and that word occurrences are weighted using weights in equation (6):

$$P(t_k|c_j, s_m) = \begin{cases} \frac{N(t_k, c_j, s_m)}{V_{c_j} + \sum_i N(t_i, c_j, s_m)} & \text{if } N(t_k, c_j, s_m) \neq 0 \\ \frac{V}{V_{c_j} + \sum_i N(t_i, c_j, s_m)} \frac{1}{V - V_{c_j}} & \text{if } N(t_k, c_j, s_m) = 0 \end{cases} \qquad (8)$$

where $N(t_k, c_j, s_m)$ is the count of the weighted occurrences of the synset t_k in the training data for class c_j in the slot s_m, V_{c_j} is the total number of unique synset in class c_j, and V is the total number of unique words across all classes. $N(t_k, c_j, s_m)$ is computed as follows:

$$N(t_k, c_j, s_m) = \sum_{i=1}^{|TR|} w_j^i n_{kim} \qquad (9)$$

In (9), n_{kim} is the number of occurrences of the term t_k in the slot s_m of the ith instance. The sum of all $N(t_k, c_j, s_m)$ in the denominator of equation (8) denotes the total weighted length of the slot s_m in the class c_j. In other words, $\hat{P}(t_k|c_j, s_m)$ is estimated as a ratio between the weighted occurrences of the synset t_k in slot s_m of class c_j and the total weighted length of the slot. The final outcome of the learning process is a probabilistic model used to classify a new instance in the class c_+ or c_-. The model can be used to build a personal profile that includes those synsets that turn out to be most indicative of the user's preferences, according to the value of the conditional probabilities in (8).

7   Experimental Evaluation

The goal of the experimental session is to evaluate whether the synset-based profiles learned by ITR actually improves the performance compared to keyword-based profiles. The evaluation was performed in the task of recommending scientific papers on the basis of the research interests stored in the user profiles. Experiments in a movie recommending domain are reported in [23]. Synset-based profiles have been also evaluated in a content-collaborative recommender system [5].

7.1   The ISWC Dataset

The ISWC dataset is a corpus of 100 papers presented during the 2002 and 2003 editions of the International Semantic Web Conference (42 papers and 58 papers respectively). Papers are rated by 11 real users on a 5-point scale mapped linearly to the interval [0,1] (see formula (6) in section 6). The dataset is described in Table 4.

Table 4. The ISWC dataset used in the experiments

Id user	Rated Papers	% POS	% NEG	n. words	n. synsets
1	37	59	41	2,702	2,546
2	22	54	46	1,597	1,506
3	27	63	37	1,929	1,792
4	27	44	56	1,830	1,670
5	29	59	41	2,019	1,896
6	22	82	18	1,554	1,433
7	26	58	42	1,734	1,611
8	28	61	39	2,034	1,901
9	23	57	43	1,442	1,374
10	22	59	41	1,335	1,258
11	25	48	52	1,740	1,640
	288	59	41	20,016	18,627

Tokenization, stopword elimination and stemming have been applied to the text in each slot in order to obtain the BOW. The content of slot *Authors* was only tokenized because it contained proper names. Documents have been processed by JIGSAW and indexed according the BOS model, obtaining a 14% feature reduction (20, 016 words vs. 18, 627 synsets), mainly due to the fact that synonym words are represented by the same synset.

7.2   Performance Measures

As our content-based profiling system is conceived as a text classifier, its effectiveness is mainly evaluated by the well-known classification accuracy measures

Table 5. Performance of the BOW - BOS profiles

	Precision		Recall		NDPM	
Id User	BOW	BOS	BOW	BOS	BOW	BOS
1	0.57	0.55	0.47	0.50	0.60	0.56
2	0.73	0.55	0.70	0.83	0.43	0.46
3	0.60	0.57	0.35	0.35	0.55	0.59
4	0.60	0.53	0.30	0.43	0.47	0.47
5	0.58	0.67	0.65	0.53	0.39	0.59
6	0.93	0.96	0.83	0.83	0.46	0.36
7	0.55	0.90	0.60	0.60	0.45	0.48
8	0.74	0.65	0.63	0.62	0.37	0.33
9	0.60	0.54	0.63	0.73	0.31	0.27
10	0.50	0.70	0.37	0.50	0.51	0.48
11	0.55	0.45	0.83	0.70	0.38	0.33
Mean	0.63	0.64	0.58	0.60	0.45	0.45

precision and *recall* [22]. Also used is F1 measure, a combination of precision and recall. We adopted the Normalized Distance-based Performance Measure (NDPM) Yao [25] to compare the ranking set by the user ratings with the ranking set by the classification scores given by ITR (the a-posteriori probability of the class *likes*). For each pair of items (d_i, d_j) in the system's ranking, a "distance score" is computed depending on whether they appear in the same order as in the user's ranking. The NDPM value is computed by averaging the "distance score" over all the possible pairs in both rankings. Values range from 0 (agreement) to 1 (disagreement). The adoption of both classification accuracy and rank accuracy metrics gives us the possibility to evaluate both whether the system is able to recommend good items and how these items are ranked. In all the experiments, a document d_i is considered as *relevant* by a user if $w_+^i > 0.5$. ITR considers an item as relevant if the a-posteriori probability of the class *likes* is greater than 0.5.

7.3   Experiment Setup and Results

We executed one run of the experiment for each user. Each run consisted in:

1. Selecting the ratings of the user and the documents rated by that user;
2. Splitting the selected data into training set *Tr* and test set *Ts* by using 5-fold cross validation;
3. Learning the user profile from *Tr*;
4. Evaluating the predictive accuracy of the induced profile on *Ts*, using the aforementioned measures.

From the results reported in Table 5, we notice an improvement both in precision (+1%) and recall (+2%). Precision improves for 4 users out of 11, while a more significant improvement (8 users out of 11) is obtained for recall.

The BOS model outperforms the BOW one specifically for users 7 and 10, for whom we observe an increased precision, and in the worst case the same recall. The rating style of these users has been thoroughly analyzed, and we observed that they provided a well balanced number of positive and negative ratings (positive examples not exceeding 60% of the training set). Moreover, they had a very clean rating style, that is, they tend to assign the score 1 to not interesting papers, and the score 5 to interesting ones.

We also observed the effect of the WSD on the training set of these users. We interpreted this effect as follows: If a polysemous word occurs both in positive and negative examples, the system is unlikely to be able to detect the discriminatory power of that feature for the classification, because the conditional probabilities of the word are almost the same for the two classes (likes and dislikes). On the other hand, once the system assigned the correct sense to the ambiguous word in each training example in which it occurred, it will be able to distinguish among the different meanings with which that word was differently used in positive and negative examples. Therefore, the occurrences of the *different* synsets assigned to the word will be heavily weighted due to the clean rating style of the users and this should result in more precise probability estimates that positively influenced the precision of the classification. By the way, the main outcome of the experiments is that it is difficult to reach a strong improvement both in precision and recall by using the BOS model: we observed a general improvement of both measures only on user 10. It could be noted from the NDPM values that the relevant/not relevant classification is improved without improving the ranking. A possible explanation of this result is that the BOS document representation has improved the classification of items whose scores (and ratings) are close to the relevant / not relevant threshold, thus the two rankings are very similar. A Wilcoxon signed ranked test, requiring a significance level $p < 0.05$, has been performed in order to validate these results. Each user is a single trial for the test. The test confirmed that there is a statistically significant difference in favor of the BOS model only as regards recall.

8   Conclusions and Future Work

We presented a system exploiting a Bayesian learning method to induce *semantic* user profiles from documents represented using WordNet synsets obtained by a WSD procedure called JIGSAW. Our hypothesis is that replacing words with synsets in the indexing phase produces a more accurate document representation that could be successfully used by learning algorithms to infer more accurate user profiles. Semantic profiles are used in the task of scientific paper recommending. Our hypothesis is confirmed by the experiments conducted in order to evaluate the effectiveness of the proposed approach and can be explained by the fact that synset-based classification allows the preference of documents with a high degree of semantic coherence, not guaranteed in case of word-based classification. As a future work, we plan to exploit not only the WordNet hierarchy but also domain ontologies in order to realize a more powerful document indexing.

References

[1] Banerjee, S., Pedersen, T.: An adapted lesk algorithm for word sense disambiguation using wordnet. In: Gelbukh, A. (ed.) CICLing 2002. LNCS, vol. 2276, pp. 136–145. Springer, Heidelberg (2002)

[2] Bloedhorn, S., Hotho, A.: Boosting for text classification with semantic features. In: Proceedings of 10th ACM SIGKDD International Conference on Knowledge Discovery and Data Mining, Mining for and from the Semantic Web Workshop, pp. 70–87. ACM Press, New York (2004)

[3] Decadt, B., Hoste, V., Daelemans, W., Van den Bosch, A.: Gambl, genetic algorithm optimization of memory-based wsd. In: Senseval-3: Third International Workshop on the Evaluation of Systems for the Semantic Analysis of Text (2002)

[4] Deerwester, S., Dumais, S.T., Furnas, G.W., Landauer, T.K., Harshman, R.: Indexing by latent semantic analysis. Journal of the American Society for Information Science 41(6), 391–407 (1990)

[5] Degemmis, M., Lops, P., Semeraro, G.: A content-collaborative recommender that exploits wordnet-based user profiles for neighborhood formation. User Modeling and User-Adapted Interaction. The journal of Personalisation Resarch (UMUAI) 17(3), 217–255 (2007)

[6] Fellbaum, C.: WordNet: An Electronic Lexical Database. MIT Press, Cambridge (1998)

[7] Guarino, N., Masolo, C., Vetere, G.: Content-based access to the web. IEEE Intelligent Systems 14(3), 70–80 (1999)

[8] Leacock, C., Chodorow, M.: Combining local context and WordNet similarity for word sense identification. In: Fellbaum, C. (ed.), pp. 305–332. MIT Press, Cambridge (1998)

[9] Lesk, M.: Automatic sense disambiguation using machine readable dictionaries: how to tell a pine cone from an ice cream cone. In: Proceedings of the 1986 SIGDOC Conference, pp. 20–29 (1986)

[10] Magnini, B., Strapparava, C.: Improving user modelling with content-based techniques. In: Proc. 8th Int. Conf. User Modeling, pp. 74–83. Springer, Heidelberg (2001)

[11] Manning, C., Schütze, H.: Foundations of Statistical Natural Language Processing. In: Text Categorization, ch. 16, pp. 575–608. The MIT Press, Cambridge (1999)

[12] Mavroeidis, D., Tsatsaronis, G., Vazirgiannis, M., Theobald, M., Weikum, G.: Word sense disambiguation for exploiting hierarchical thesauri in text classification. In: Jorge, A.M., Torgo, L., Brazdil, P.B., Camacho, R., Gama, J. (eds.) PKDD 2005. LNCS (LNAI), vol. 3721, pp. 181–192. Springer, Heidelberg (2005)

[13] McCallum, A., Nigam, K.: A comparison of event models for naive bayes text classification. In: Proceedings of the AAAI/ICML-98 Workshop on Learning for Text Categorization, pp. 41–48. AAAI Press, Stanford (1998)

[14] Mihalcea, R.: Unsupervised large-vocabulary word sense disambiguation with graph-based algorithms for sequence data labeling. In: Proceedings of the Joint Conference on Human Language Technology / Empirical Methods in Natural Language Processing (HLT/EMNLP) (2005)

[15] Mihalcea, R., Chklovski, T.: Open Mind Word Expert: Creating Large Annotated Data Collections with Web Users' Help. In: Proceedings of the EACL Workshop on Linguistically Annotated Corpora, Budapest (2003)

[16] Mihalcea, R., Csomai, A.: Senselearner: Word sense disambiguation for all words in unrestricted text. In: Proceedings of the 43rd Annual Meeting of the Association for Computational Linguistics (2005)

[17] Miller, G., Beckwith, R., Fellbaum, C., Gross, D., Miller, K.: Introduction to Wordnet: an on-line lexical database. International Journal of Lexicography (Special Issue) 3(4), 235–244 (1990)

[18] Mooney, R.J., Roy, L.: Content-based book recommending using learning for text categorization. In: Proceedings of the 5th ACM Conference on Digital Libraries, San Antonio, US, pp. 195–204. ACM Press, New York (2000)

[19] Navigli, R., Velardi, P.: Structural semantic interconnections: A knowledge-based approach to word sense disambiguation. IEEE Transactions on Pattern Analysis and Machine Intelligence 27(7), 1075–1086 (2005)

[20] Resnik, P.: Disambiguating noun groupings with respect to WordNet senses. In: Proceedings of the Third Workshop on Very Large Corpora, pp. 54–68. Association for Computational Linguistics (1995)

[21] Scott, S., Matwin, S.: Text classification using wordnet hypernyms. In: COLING-ACL Workshop on usage of WordNet in NLP Systems, pp. 45–51 (1998)

[22] Sebastiani, F.: Machine learning in automated text categorization. ACM Computing Surveys 34(1), 1–47 (2002)

[23] Semeraro, G., Degemmis, M., Lops, P., Basile, P.: Combining Learning and Word Sense Disambiguation for Intelligent User Profiling. In: Proceedings of the Twentieth International Joint Conference on Artificial Intelligence, Hyderabad, India, January 6-12, 2007, pp. 2856–2861. Morgan Kaufmann, San Francisco (2007)

[24] Witten, I.H., Bell, T.C.: The zero-frequency problem: Estimating the probabilities of novel events in adaptive text compression. IEEE Transactions on Information Theory 37(4), 1085–1094 (1991)

[25] Yao, Y.Y.: Measuring retrieval effectiveness based on user preference of documents. Journal of the American Society for Information Science 46(2), 133–145 (1995)

[26] Yuret, D.: Some experiments with a naive bayes wsd system. In: Senseval-3: Third International Workshop on the Evaluation of Systems for the Semantic Analysis of Text (2002)

Web Usage Mining in Noisy and Ambiguous Environments: Exploring the Role of Concept Hierarchies, Compression, and Robust User Profiles

Olfa Nasraoui and Esin Saka

Knowledge Discovery & Web Mining Lab,
University of Louisville, Louisville, KY 40292, USA
http://webmining.spd.louisville.edu

Abstract. Recent efforts in Web usage mining have started incorporating more semantics into the data in order to obtain a representation deeper than shallow clicks. In this paper, we review these approaches, and examine the incorporation of simple cues from a website hierarchy in order to relate clickstream events that would otherwise seem unrelated, and thus perform URL compression. We study their effect on data reduction and on the quality of the resulting knowledge discovery. Web usage data is also notorious for containing moderate to high amounts of noise, thus motivating the use of robust knowledge discovery algorithms that can resist noise and outliers with various degrees of resistance or robustness. Therefore, we also examine the effect of robustness on the final quality of the knowledge discovery. Our experimental results conclude that post-processed and robust user profiles have better quality than raw profiles that are estimated through optimization alone. However URL compression, as expected, tends to reduce the quality, but also can drastically reduce the size of the data set, resulting in faster mining.

1 Introduction

Mining Web server access log data or *Web usage mining* offers some of the most promising techniques to analyze data generated on a Website in order to help understand how users navigate through a given website, what information appeals to their interests, and what peculiar information needs drive them in their browsing sessions. In addition to this understanding, Web usage mining can be used to improve a Website design and to provide automated and intelligent personalization that tailors a user's interaction with the website based on the user's interests. Understanding Web users' browsing patterns and personalizing their web navigation experience is beneficial to all users, but it is particularly crucial on websites that are visited by a large variety of visitors with varying levels of expertise.

Traditionally, Web usage data has been represented as a bag or a sequence of clicks or URLs that are visited by a user during a single visit to a website.

B. Berendt et al. (Eds.): WebMine 2006, LNAI 4737, pp. 82–101, 2007.

However, earlier efforts toward mining Web usage data have been limited by the shallowness of representation at the "click" level. This has prompted later efforts to incorporate content and more recently, even semantics to obtain a deeper representation. In this paper, we review these approaches, and examine the incorporation of simple cues from a website hierarchy in order to relate clickstream events that would otherwise seem unrelated, and study their effect on data reduction and on the quality of the resulting knowledge discovery. Web usage data is also notorious for containing moderate to high amounts of noise, thus motivating the use of robust knowledge discovery algorithms that can resist noise and outliers with various degrees of resistance or robustness. Therefore, we also examine the effect of robustness on the final quality of the knowledge discovery. Our experimental results conclude that post-processed and robust user profiles have better quality than raw profiles that are estimated through optimization alone. However URL compression, as expected, tends to reduce the quality, but also can drastically reduce the size of the data set, resulting in faster mining.

2   Related Work

2.1   Mining Mass User Profiles from Web Clickstreams

Personalization deals with tailoring a user's interaction with the Web information space based on information about him/her, in the same way that a reference librarian uses background knowledge about a person or *context* in order to help them better. The concept of *contexts* can be mapped to distinct user *profiles*. Manually entered profiles have raised serious *privacy* concerns, are *subjective*, and *do not adapt* to the users' changing interests. *Mass profiling*, on the other hand, is based on general trends of usage patterns (thus protecting privacy) compiled from all users on a site, and can be achieved by mining or discovering user profiles (i.e., clusters of similar user access patterns) from the historical *web clickstream* data stored in server access logs. The simplest type of personalization system can suggest relevant URLs or links to a user based on the user's interest as inferred from their recent URL requests. A *web clickstream* is a virtual trail that a user leaves behind while surfing the Internet, such as a record of every page of a Web site that the user visits. Recently, data mining techniques have been applied to discover mass usage patterns or profiles from Web log data [36,29,20,7,28,14,1,17,5,33,30,32,34,31,35,4]. Most of these efforts have proposed using various data mining or machine learning techniques to model and understand Web user activity. In [34], clustering was used to segment user sessions into clusters or profiles that can later form the basis for personalization. In [29], the notion of an adaptive website was proposed, where the user's access pattern can be used to automatically synthesize index pages. the work in [8] is based on using association rule discovery as the basis for modeling web user activity, while the approach proposed in [4] used Markov Random Fields to model Web navigation patterns for the purpose of prediction. The work in [36] proposed building data cubes from Web log data, and later applying Online Analytical

Processing (OLAP) and data mining on the cube model. [31] presented a complete Web Usage Mining (WUM) system that extracted patterns from Web log data with a variety of data mining techniques. In [23,24], we have proposed new robust and fuzzy relational clustering techniques that allow Web usage clusters to overlap, and that can detect and handle outliers in the data set. A new subjective similarity measure between two Web sessions, that captures the organization of a Web site, was also presented as well as a new mathematical model for "robust" Web user profiles [24]. In [20], a *linear* complexity Evolutionary Computation technique, called Hierarchical Unsupervised Niche Clustering (H-UNC), was presented for mining both user profile clusters and URL associations. A density based evolutionary clustering technique is proposed to discover multi-resolution and robust user profiles in [22]. The K Means algorithm was used in [30] to segment user sequences into different clusters. An extensive survey of different approaches to Web usage mining can be found in [32].

2.2   Semantics and Concept Hierarchies

Relying only on Web usage data for personalization can be inefficient either when there is insufficient usage data for the purpose of mining certain patterns, or when new pages are added and thus do not accumulate sufficient usage data at first. Lack of usage data in these cases can be compensated by adding other information such as the content of Web pages [18] or the structure of a Web site [24,23]. In [18], the keywords that appear in web pages are used to generate document vectors, which are later clustered in the document space to further augment user profiles. In [24,23], the website's own hierarchical structure is treated like an *implicit* taxonomy or concept hierarchy that is exploited in computing the similarity between any two web pages on the website. This allows a better comparison between sessions that contain visits to web pages that are different, and yet semantically related (for example under the same more general topic). The idea of exploiting concept hierarchies or taxonomies has already been found to enhance association rule mining in [2] and to facilitate information searching in textual data [6]. For example, our similarity measures initially proposed in [24,23] have lately been generalized in [11] to the context of digital libraries which often benefit from an implicit taxonomy. Even though keywords present in the Web pages have been used to add a content aspect to usage data, the keyword based approach remains incapable of capturing more complex relationships at a deeper semantic level based on different types of attributes associated with structured objects. In [9], a general framework was proposed for using domain ontologies to automatically characterize usage profiles containing a set of structured Web objects. This framework allows a Web personalization that goes beyond the low level of raw pages or items, and instead exploits the semantic power of an underlying ontology.

The advent of dynamic URLs mostly in tandem with Web databases has recently made it even more difficult to interpret URLs in terms of user behavior, interests, and intentions. For instance, consider the following cryptic association rule within the context of an online bookstore, which was included in an example from [27]:

If http://www.the_shop.com/show.html-?item=123,

Then http://www.the_shop.com/show.html?item=456, support= 0.05, confidence = 0.4.

A more meaningful rule would be Users who bought "Hamlet" also tended to buy "How to stop worrying and start living". This in turn has motivated the work in [27] to mine patterns of application events instead of patterns of URLs, by exploiting the semantics of the pages visited along user paths while performing Web usage mining. Within this spirit, Service based concept hierarchies were introduced in [3] for analyzing the search behavior of visitors, i.e. how they navigate rather than what they retrieve. In this case, concept hierarchies form the basic method of grouping Web pages together, and later extending Web usage mining by identifying the differences between navigation patterns, and exploiting the site's semantics in the visualization of the results. Web pages are treated as instances of a higher-level concept, based on page content or service requested. This in turn leads to abstracted pages or paths. In [27], usage mining was enhanced by registering the user behavior in terms of an ontology underlying a particular website. The semantic annotation of the Web content is assumed to have been performed a priori, since the website in question is a knowledge portal with an inherent RDF annotation. In order to mine interesting patterns, first, the Web logs are semantically enriched with ontology concepts. Then, these semantic Web logs are mined to extract patterns such as groups of users, users preferences and rules. Following a similar approach, in [10] Web usage logs were enriched with semantics derived from the content of the Web site's pages. The enhanced Web logs, called C-Logs are then used as input to the Web mining process, resulting in the creation of a broader set of recommendations. First, the extraction of the keywords that describe each Web page is performed using information retrieval based techniques. These keywords are then mapped to the categories of a predefined domain-specific taxonomy through the use of a thesaurus. The taxonomy is constructed manually by a domain expert. Following keyword extraction, the Web documents are clustered based on the taxonomy categories, and the recommended categories are further expanded to contain the documents that fall under the same category.

We should note that with the exception of [24,23] that use an implicit taxonomy to relate web pages, most other efforts cited above, rely on an explicit taxonomy. Explicit taxonomies, in all these efforts, need to be hand-crafted by an expert before the analysis. The implicit taxonomy in [24,23], on the other hand, is inferred automatically and quickly from the website directory structure via URL tokenization. Furthermore, this implicit taxonomy does not require any modification to the underlying data mining algorithm, since it is incorporated and isolated only within the similarity measure used to compare sessions. In this paper we will exploit an *implicit* taxonomy as inferred from the website directory structure.

2.3   Noise and Robustness

The problem of the contamination of the input data by noise or outliers is faced by most data mining applications. In the case of user sessions this means that some of the input sessions will either be incorrect as a result of the session construction and unique session identification process in pre-processing, or that some user sessions are naturally outlying relative to the majority of the sessions, i.e. they differ from most data. By definition, noise tends to be different from and typically occurs less frequently than the majority of the good data. This problem is more acute in the presence of Power law distributions [16] because of of the presence of a long tail consisting of infrequent items, such that the aggregate frequencies of all the long tail items constitutes the majority of the items! This means that significant proportions of the user sessions can be expected to be noisy. This severe noise contamination requires robust clustering techniques that can discover meaningful clusters consisting of similar user sessions, while resisting the effect of noise. This is one reason why we relied on H-UNC to cluster the user sessions. One way that H-UNC handles noise is by the automated estimation of robust weights that are low for noise and outliers and high for clean or good data, and by relying on these robust weights to estimate a robust measure of scale and robust measure of density in each cluster. These robust weights allow the identification and elimination of noise sessions in each discovered cluster both while searching for the optimal clusters (thus improving the quality of clustering), and later while post-processing the clusters to extract *robust user profiles*.

3   Exploiting Concept Hierarchies

3.1   Overview of the KDD Process with Hierarchical Unsupervised Niche Clustering

The framework for our web usage mining starts with the collection of Web server logs, follows with standard pre-processing, such as data cleaning and sessionization, then continues with the pattern discovery via clustering, and ends with the post-processing, interpretation, and evaluation of the discovered user profiles. The final user profiles are intended to be later used to recommend relevant URLs to new anonymous users of a Web site. The knowledge discovery part can be executed offline by periodically mining new contents of the user access log files, and can be summarized in the following steps:

1. Preprocess log file to extract user sessions,
2. Cluster user sessions by Hierarchical Unsupervised Niche Clustering (H-UNC) [22,20]
3. Post-process the clusters of sessions into summary and robust user profiles,
4. Evaluate the user profiles.

Step 1: Preprocessing the Web Log File to extract User Sessions: The access log of a Web server is a record of all files (URLs) accessed by users on a Web site.

Each log entry consists of the following information components: access time, IP address, URL viewed, etc. The first step in preprocessing [8,25] consists of mapping the N_U URLs on a website to distinct indices. A user session consists of requests originating from the same IP address within a predefined time period. Each URL in the site is assigned a unique number $j = 1, \ldots, N_U$, where N_U is the total number of valid URLs. Thus, the jth user session is encoded as an N_U-dimensional binary attribute vector s_j with the property

$$s_{jk} = \left\{ \begin{array}{c} 1 \text{ if } URL_k \in \text{ session } j \\ 0 \text{ otherwise} \end{array} \right\}.$$

We also take advantage of the Power law properties of Web session lengths [16]. Therefore, even though sessions are mathematically modeled as binary vectors, they are implemented as lists to save on memory and computational costs, since most sessions are very short compared to the total number of URLs.

Step 2: Clustering Sessions into an Optimal Number of Categories: For this task, we use Hierarchical Unsupervised Niche Clustering [22] or H-UNC. H-UNC is a hierarchical version of a robust genetic clustering approach (UNC) [21]. Inspired by nature, a Genetic Algorithm (GA) [13] evolves a population of candidate solutions through generations of competition and reproduction until convergence to one solution. Hence, the GA cannot maintain population diversity. Niching methods, on the other hand, attempt to maintain a diverse population in a GA with members distributed among niches corresponding to multiple solutions. An initial population of randomly selected sessions is encoded into binary chromosome strings that compete based on a density based fitness measure that is highest at the centers of good (dense) clusters. Different niches in the fitness landscape correspond to distinct clusters in the data set. The main outline of the H-UNC algorithm is sketched below. The reason why we use H-UNC instead of other clustering algorithms is that unlike most other algorithms, H-UNC allows the clusters and their number to be determined automatically in a gradual multi-resolution process, and the noise and outliers to be tolerated. Furthermore, the evolutionary optimization in H-UNC gives us the freedom to cluster the data using any similarity measure, particularly, a subjective measure that exploits domain knowledge, such as given below in (1). Also, unlike purely evolutionary search based algorithms, H-UNC is a *hybrid* technique that combines fast local Piccard updates to estimate the scale or variance, σ_i, of the data in the cluster around each candidate profile. This fast local search makes the evolutionary search converge fast (typically in 20 generations). Below we list the steps of H-UNC, while more details on H-UNC can be found in [22,20].

Step 3: Post-processing Session Clusters into Summary User Profiles: In order to post-process the raw profiles, determined by optimization alone in H-UNC, we first automatically group the sessions s_j into different clusters, $\chi_i = \{s_j \mid Dist(i,j) < Dist(k,j), \forall k \neq i\}$, based on the Web session distance, $Dist(i,j)$, defined in (5) from session x_j to the closest *raw* profile p_i^{raw}. Then, we

---

Algorithm. Hierarchical Unsupervised Niche Clustering (H-UNC)
Input: x_j : user sessions, N_{min} : minimum allowed cluster cardinality,
　　　σ_{split} : minimum allowed cluster scale
Output: p_i: User profiles (sets of URLs, later referred to as *raw* profiles)
　　　χ_i: Clusters of user sessions closest to profile i

1: 　Encode binary session vectors x_j
2: 　Set current resolution Level $L = 1$
3: 　Apply UNC (one level clustering) to entire data set with small population size
　　　　(this results in cluster representatives p_i and corresponding scales σ_i)
4: 　**repeat until** cluster cardinality $N_i < N_{min}$ or scale $\sigma_i < \sigma_{split}$
5: 　　　Increment resolution level: $L = L + 1$
6: 　　　For each parent cluster p_i found at Level $(L - 1)$
7: 　　　**if** cluster cardinality $N_i > N_{min}$ **or** cluster scale $\sigma_i > \sigma_{split}$ **then**
8: 　　　　　Reapply UNC [21] only on data records x_j that are closest to this
　　　　　　parent cluster p_i based on distance measure in (5)
　　　　　　(this results in cluster representatives p_i and corresponding scales σ_i)

Algorithm. Unsupervised Niche Clustering (UNC)
Input: x_j : data records, in this case user sessions,
　　　N_p: population size, G: Number of generations
Output: Cluster representatives: a set of profiles p_i and scales σ_i

---

1: 　Randomly select from x_j an initial population of N_p candidate representatives p_i
2: 　Set initial scales $\sigma_i = \frac{\max_{i,j} d_{ij}}{10}$
3: 　**repeat for** G generations
4: 　　　Update the distance d_{ij} of each data record x_j relative
　　　　to each candidate cluster representative using distance defined in (1)
5: 　　　Update the robust weight $w_{ij} = e^{-d_{ij}/(2\sigma_i)}$ of each data record x_j
　　　　relative to each candidate cluster representative p_i
6: 　　　Update the scale $\sigma_i = \frac{\sum_j w_{ij} d_{ij}}{\sum_j w_{ij}}$ for each candidate cluster representative
　　　　(derived by setting fitness gradient: $\partial f_i / \partial \sigma_i = 0$ while w_{ij} are fixed)
7: 　　　Update density fitness $f_i = \frac{\sum_j w_{ij}}{\sigma_i}$ of each cluster representative p_i
8: 　　　**for** $i = 1$ **to** $N_p/2$ **do**
9: 　　　　　Select randomly from population candidate parent p_i
　　　　　　without replacement
9: 　　　　　Select randomly from population another candidate parent p_k
　　　　　　without replacement
11: 　　　　　Obtain children c_1 and c_2 by performing crossover and mutation
　　　　　　between the chromosome strings of p_i and p_k
12: 　　　　　Update the scale σ_i and the fitness f_i of each child
13: 　　　　　Apply Deterministic Crowding to fill new population:
　　　　　　　　Assign each child c_i to closest parent
　　　　　　　　if child's fitness > closest parent's fitness **then**
　　　　　　　　　child replaces closest parent in the new population
　　　　　　　　else closest parent remains in the new population

summarize the session clusters in terms of *post-processed* user profile vectors p_i as proposed in [20]: The kth component/weight of this vector ($p_{ik} = \frac{\sum_{s_j \in X_i} s_{jk}}{|\chi_i|}$) captures the relevance of URL_k in the ith profile, as estimated by the conditional probability that URL_k is accessed in a session belonging to the ith cluster (this is the frequency with which URL_k was accessed in the sessions belonging to the ith cluster). The profiles p_i are then binarized so that only URLs URL_k with weights $p_{ik} > 0.15$ remain. In addition, each post-processed profile p_i inherits the scale of the corresponding raw profile that is estimated while optimizing profiles. The scale σ_i represents the amount of variance or dispersion of the user sessions in a given cluster around their closest cluster session representative.

3.2 Using the Website Hierarchy to Relate What Would Otherwise Seem Unrelated

The similarity score between two input sessions s_k and s_l can be computed as follows (where N_U is the total number of URLs) [24,23]

$$Sim_{session}(k,l) = max\{cosine(k,l), cosine_{hierarchical}(k,l)\} \tag{1}$$

where

$$cosine(k,l) = \frac{\sum_i s_{ki} s_{li}}{\sqrt{\sum_i s_{ki} \sum_i s_{li}}} \tag{2}$$

and

$$cosine_{hierarchical}(k,l) = \frac{\sum_i \sum_j s_{ki} s_{lj} S_u(i,j)}{\sum_i s_{ki} \sum_i s_{li}} \tag{3}$$

is a modification of the cosine similarity, that we introduced in [3,4], and that can take into account the Website structure in order to account for the similarities between distinct URLs or items i and j that share some conceptual similarity, given by $S_u(i,j)$, which is a URL to URL similarity matrix that is computed based on the amount of overlap between the paths leading from the root of the website (main page) to the two URLs i and j, and is given by

$$S_u(i,j) = min\left(1, \frac{|LCP(i,j)|}{max\left(1, max\left(|path(URL_i)|, |path(URL_j)|\right) - 1\right)}\right) \tag{4}$$

where $LCP(i,j)$ is the Longest Common Prefix of URLs i and j. For example, if $URL_i = a/b/c$ and $URL_j = a/b/d$, then $|path(URL_i)| = 3$, $|path(URL_j)| = 3$, and $LCP(i,j) = a/b$ which consists of 2 tokens (separated by "/"), hence its length will be $|LCP(i,j)| = 2$. We subtract 1 from the maximum of the path lengths in the denominator so that the root ("/") is not counted. We refer to the special similarity in (1) as the *Web Session Similarity*. This web similarity takes into account the hierarchical structure of website content as inferred from the URL address itself, by comparing the prefixes of two URLs. Another way to induce a concept hierarchy is to assess how different content items on the website relate to each other according to an externally defined website ontology, as we have done in [26]. This similarity is used in our clustering algorithm (H-UNC) to group similar user sessions into clusters or profiles. The URL to URL similarities in (4) form a sparse matrix, hence only non-zero values are stored. Furthermore, access to these values is accelerated by hashing the two indices corresponding to a given pair of URLs. Also for the purpose of clustering, the Web session similarity measure in (1) is mapped to a distance as follows:

$$Dist(k, l) = (1 - Sim_{session}(k, l))^2 \tag{5}$$

that measures the dissimilarity between a session and another session or a session and a cluster representative prototype or profile, since the latter has the same format as a user session. The distance is squared to enhance the contrast between small distances (close to 0) and large distances (close to 1). This contrast will later be reflected in the robust weights, and thus the density fitness of the candidate profiles. Our preliminary results showed that this improves competition between the profile candidates in the population.

In this paper we exploit an implicit taxonomy as inferred from the website directory structure or Website hierarchy, which is available as a cheap and available source of information. One can also use an explicit taxonomy as inferred from external taxonomy data if available. For example, some dynamic URLs can be related through a concept hierarchy that can be decoded from the URLs. Additional information about a catalog of items of the content being served can be used to map each content item or URL to a tokenized URL string We have previously explored this explicit URL hierarchy approach in [26]. The implicit or explicit taxonomy information are seamlessly incorporated into the data mining algorithm (in this case clustering) via the computation of the special session similarity measures (4) and (1).

3.3   Using the URL Similarities for Data Compression

For a large website, the number of URLs may be very large. In this case, we can identify which URLs are so similar, that they may be merged for computational purposes, using the URL to URL similarity S_u in (4). We accomplish the compression by recursively merging URLs that have similarity $S_u > S_{min}$ into one URL. Essentially, this amounts to (1) forming a graph $\mathcal{G}$ with a vertex for each URL, and with an edge connecting two URLs only if they have high similarity, i.e. $S_u > S_{min}$, then (2) partitioning $\mathcal{G}$ into maximally connected components, and finally (3) merging the URLs in each component as one pseudo-URL. Naturally, higher thresholds S_{min} will result in sparser or less connected graphs, and thus more clusters and less compression. Recursive clustering performs a transitive closure on the similarity if it is considered as a fuzzy relation matrix [15], thus effectively transforming the corresponding distance into an Euclidean space. On websites that have a very deep structure (both directory and semantic), it is possible to gain tremendous savings in the number of URLs used for data mining. For example, it is typical for the number of URLs to be reduced to

90% of the original number of URLs, meaning that a compression ratio of 90% is achieved.

3.4   Effects on Item Abundance and Data Sparsity

The distribution of URL access frequencies is known to follow a Zipf power law [16], which therefore exhibits a long tail that consists of the majority of the URLs. These long tail majority URLs are however accessed very infrequently. It can be shown that compression can have a tremendous effect on the size of the long tail and on the sparsity of the user session matrix. Given a hierarchically structured website or any concept hierarchy with a large number of items (in our case, URLs), and in order to quantify its size, assume that the average branching factor of this tree is b, and that its depth or number of levels is D. Then the number of URLs at the leaves is at most $N_U = b^D$. If compression is performed by merging all URLs that share a common prefix path up to D' levels with D' being a Cth fraction of the original full depth D, i.e., $D' = D/C$, then the maximum number of URLs at the leaves is reduced to $N'_U = b^{D'} = b^{D/C}$. Since $N_U = b^D = b^{(D/C)C} = \left(b^{(D/C)}\right)^C = (N'_U)^C$, it follows that the compressed number of URLs $N'_U = (N_U)^{1/C}$, which can be significantly lower than the original number of URLs N_U.

Since compression has the effect of rolling up the concept hierarchy so that the depth of the hierarchy decreases from D to D', the probability of an item or URL at the leaves being accessed in any given session will also increase from $\delta = P\{\text{item at level } D\}$ in the original uncompressed space to $\delta_c(i) = P\{\text{item i at level } D'\}$ in the compressed space, where $\delta_c(i) = \sum_{k\in descendants(i)} \delta_k$ since each item/URL at level D' is an ancestor of all items k at level D, in the compressed URL hierarchy, that are its descendants. Assuming a branching factor b and assuming a compression ratio $C = D/D'$, each item at level D' will have on average $b^{(D-D')} = b^{(D-D/C)} = b^{D(1-1/C)}$ descendants. Therefore the average density after compression will be approximately $\delta_c \approx b^{D(1-1/C)}\delta$. Hence the density increase ratio is $\delta_c/\delta \approx b^{D(1-1/C)}$, which increases with the branching factor b, the URL hierarchy depth D, and the depth reduction factor C which affects the strength of compression. As item density increases, the *sparsity* of the sessions $(1 - \delta)$ also decreases at the same rate. This means that sparsity can be expected to decrease as a consequence of URL compression, and that the decrease in sparsity will be more significant for website hierarchies that are deeper and with more branches.

4   Robustness to Noise

4.1   Characterizing Noise Sessions and Robust Core Profiles

As it estimates the cluster profiles, H-UNC computes a robust weight $w_{ij} = e^{-Dist(i,j)/(2\sigma_i)}$ for each user session s_j relative to the closest cluster representative p_i based on the Web session distance measure in (5). Based on these weights, we can easily distinguish between *core* sessions that are well represented by their cluster profile, and thus receive a high weight, and *noise* sessions that receive low weight because they are far from the cluster profile. The *core* of a profile is defined as $\chi_i^{core} = \{s_j \in \chi_i \mid w_{ij} > W_{min}\}$. As in the case of the original cluster, we summarize the session in each core in terms of *a robust* user profile vector [23], p_i^{robust}. The kth component/weight of this vector is

$$p_{ik}^{robust} = \frac{\sum_{s_j \in \chi_i^{core}} s_{jk}}{|\chi_i^{core}|} \tag{6}$$

Unpolluted by noisy and irrelevant sessions, the robust profiles give a cleaner description of the user interests in each cluster of user sessions.

4.2   Effects of Post-processing on Precision and Coverage of User Profiles

The H-UNC algorithm starts with a pool of candidate user profiles that get picked randomly from the input sessions. Because most user sessions tend to be short compared to the full range of URLs, these candidate profiles are short, and because most similar sessions that form a good cluster also tend to be short, the final optimized profiles tend to converge toward the most typical (frequent) sessions, and therefore the optimized profiles are also short. This is because the fitness f_i of a candidate profile p_i in H-UNC is a measure of density (the ratio of the sum of robust weights $\sum_{s_j \in \chi_i^{core}} w_{ij}$ to the scale of the cluster σ_i) of the user sessions around the candidate profile. These profiles determined solely by optimization are called the *raw profiles*. Post-processing results in profiles that have URL weights that are averages of frequencies of access to each URL in the sessions assigned (closest) to this profile. This averaging of the frequencies of URL accesses in each cluster will usually preserve the original URLs of the raw (unprocessed) optimized profile (because they are already frequent), but will typically bring more URLs into the profile (as long as they are accessed in more than 15% of the sessions in that cluster). This emergence of more URLs can be expected to enhance recall.

The effect of robust post-processing on precision and recall can be analyzed as follows. As the minimum robust weight threshold W_{min} increases, more of the noise sessions are *ignored* from the computation of the aggregate user profiles, thus reducing the denominator in (6), which is simply the count of data records that are assigned to the cluster. On the other hand, its numerator is not affected as much, because outliers (which by definition have low similarity with the cluster sessions) do not share any significant components (i.e. URLs) with the cluster profile, and thus outliers do not contribute to the numerator (only good data records contribute). Thus, the decrease in the denominator of (6) can be expected to increase the number of good URLs that remain (with sufficient frequency) in the final robust profile. The effect of bringing more URLs into the robust profile can therefore be expected to increase recall and possibly slightly reduce precision depending on the robust weight threshold W_{min}.

5   Experimental Results

Hierarchical Unsupervised Niche Clustering (H-UNC) [22] was applied on a set
of web sessions preprocessed from real Web log data, as follows[25]: After filtering out irrelevant entries, the data was segmented into unique sessions based
on the client IP address and a timeout threshold: The maximum elapsed time
between two consecutive accesses in the same session was set to 45 minutes. The
results of pre-processing two Web logs were two session data sets consisting of
1,704 sessions and 343 URLs for a computer science and engineering department
website, and 11,036 sessions and 8,656 URLs for a university library website. H-
UNC was applied to the Web sessions using a maximal number of levels $L = 3$
in the recursive clustering, and the following parameters that control the final
resolution [22]: $N_{split} = 30$ and $\sigma_{split} = 0.01$. H-UNC partitioned the Web user
sessions into several clusters at level 3, and each cluster was characterized by
one of the profile vectors, p_i. We will try to quantify the effect of varying the
URL compression strength on the final quality of the user profiles by first merging compatible URLs based on varying thresholds in the URL similarities, and
then performing H-UNC clustering to discover user profiles. Then, we will study
the effect of post-processing, which is an optional step that follows H-UNC, and
finally study the effect of varying the robustness level on the quality of the post-
processed user profiles. In order to assess the quality of the user profiles, we first
explain our profile validation metrics, that we have proposed in [19], in the next
section, and then present our results.

5.1   Validation Metrics

We can view the discovered profiles as frequent patterns that provide one way to
form a summary of the input data. As a summary, profiles represent a reduced
form of the data that is at the same time, as close as possible to the original
input data. This description is reminiscent of an information retrieval scenario,
in the sense that profiles that are retrieved as a result of a data query (an input
session) should be as close as possible to the original session data. Closeness
should take into account both *(i)* precision (a summary profile's items are all
correct or included in the original input data, i.e. they include only the true data
items) and *(ii)* coverage/recall (a summary profile's items are complete compared
to the data that is summarized, i.e. they include all the data items). These
criteria are clearly contradictory, since precision will favor only the smallest
profiles, eventually with a single URL, while coverage will favor the longest
possible profiles. Ideally, for perfect retrieval, each data query should be answered
by a profile that is identical to this query. However, this is unrealistic since it
corresponds to the case where the profiles summary is identical to the entire
input database. Therefore, it is imperative that the summary consist of the
smallest number of profiles that are as similar as possible to the input data. We
propose a validation procedure that attempts to answer the following crucial
questions [19]:

(a) Is the data set completely summarized/represented by the mined profiles?
(b) Is the data set faithfully/accurately summarized/represented by the mined profiles?

Each of the previous questions is answered by computing coverage/recall as part of a quality or interestingness measure to answer part (a), and precision as part of a quality/interestingness measure to answer part (b). First, we compute the following Interestingness measures for each discovered profile, letting the Quality or interestingness measure, $Q_{ij} = Cov_{ij}$ (i.e., coverage) to answer part (a), and $Q_{ij} = Prec_{ij}$ (i.e., precision) to answer part (b), where coverage and precision for a discovered mass or cluster profile p_i as a summary of an input session s_j, are given by $Cov_{ij} = \frac{|s_j \cap p_i|}{|s_j|}$ and $Prec_{ij} = \frac{|s_j \cap p_i|}{|p_i|}$. A combined measure of precision and coverage is given by the F_1 information retrieval metric, $Q_{ij} = F_{1_{ij}}$, which answers (a) and (b) simultaneously, and is given by $F_{1_{ij}} = 2Prec_{ij}Cov_{ij}/(Prec_{ij} + Cov_{ij})$. If we let $S^* = \{s_j \in S \mid max_i(Q_{ij}) \geq Q_{min}\}$ be the subset of input user sessions S that are summarized by any of the user profiles p_i with quality level higher than a given minimum quality threshold Q_{min}, then the overall quality of an entire *set* of discovered user profiles p_i can be measured by [19]

$$Q = \frac{|S^*|}{|S|}$$

where $|.|$ denotes the *cardinality* of a set. When $Q_{ij} = Cov_{ij}$, we call Q the *Cumulative Coverage of sessions*, and it answers Question (a). When $Q_{ij} = Prec_{ij}$, we call Q the *Cumulative Precision of sessions*, and it answers Question (b). *Both questions are answered when $Q_{ij} = F_{1_{ij}}$.* The quality of discovered profiles Q can also be interpreted from a probabilistic point of view as the probability that the quality with which a discovered profile summarizes *any* input session is higher than a minimum level Q_{min}. In our evaluation experiments, the above measures are computed over the entire range of the quality threshold Q_{min}, from 0% to 100% in increments of 10%, and compared, as shown in Figure 1. This validation process tries to predict, in advance, how the discovered profiles would fare if used as part of a nearest profile collaborative filtering recommender system, assuming that the user sessions remain similar to the sessions used to mine the user profiles. This validation strategy can serve *(i) to evaluate the effect of URL compression* at different URL similarity thresholds, and *(ii) to compare various profile post-processing schemes*, including the effect of robust profiles. Finally, it is important to note that our evaluation targets the quality of the mass user profiles as a summary of the input sessions, and is not necessarily relevant within the framework of a recommendation system. In other words, what we are interested in, is a fast pre-diagnostic evaluation of the quality of the data mining task, and not the quality of personalization, which after all, will strongly depend on the eventual personalization strategy that would be implemented based on the discovered user profiles. Our validation should be seen as more analogous with the validation of the results of clustering algorithms [12].

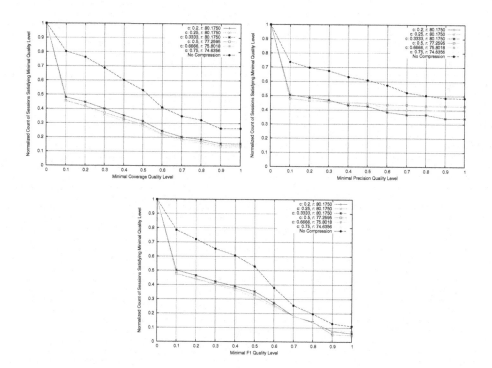

Fig. 1. Effect of different compression levels on the achieved cumulative quality Q (on y-axis) measured in terms of (from left to right): coverage, precision, F_1, versus the minimum quality threshold Q_{min}(on x-axis). Legend explanation: c : minimum URL similarity threshold S_{min} used in compression, r : compression ratio (ratio of the number of reduced URLs to total number of URLs).

5.2 Effect of URL Compression

Websites with a very high number of URLs tend to generate massive usage data. In this case, we can merge similar URLs for computational purposes, using the URL to URL similarity in (4) which takes into account the site structure overlap between the URLs. We accomplish the compression by merging URLs that have similarity $S_u > S_{min}$ into one URL, resulting in tremendous savings in the size of the data that has to undergo the data mining phase. Best of all, if meaningful similarities form the basis of the compression, then the quality of the results should not be significantly compromised as a result of compression. In fact in some cases, the quality of the discovered clusters may be helped because compression decreases the sparsity (as shown above), along with the size of the data. Figure 1 shows the results of the validation procedure using the cumulative quality metric Q, defined in (5.1), that is used to verify the quality of the discovered profiles, with and without compression, for the user input sessions extracted from the Web access logs of a computer science department server (with 1704 sessions and 343 URLs). The figures show the percentage of sessions (relative to

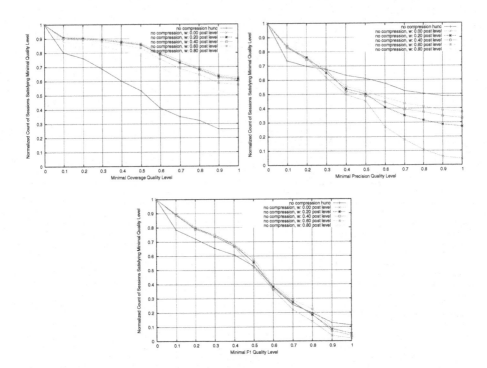

Fig. 2. Effect of different robustness levels used during post-processing on the achieved cumulative quality Q (on y-axis) measured in terms of (from left to right): coverage, precision, F_1, versus the minimum quality threshold Q_{min}(on x-axis). <u>Legend explanation:</u> *no compression hunc:* results without post-processing, all others are with post-processing at varying robustness levels, W : Minimum robust weight threshold (W_{min}) used in post-processing *robust* profiles (W:0.00 means that no noise is excluded while post-processing, i.e. not robust).

the entire Web log) which can be retrieved with a given minimal quality level by one of the discovered profiles, where the minimal quality level, Q_{min} is varied from 0 to 100%. The quality measures shown are the Precision, Coverage, and F_1. By reading the y-axis for a given minimal quality level on the x-axis, we can obtain the percentage of sessions that achieve a given quality level. For instance Figure 1 (a) shows that without compression, 54% of the sessions achieve a coverage = 0.5; while Figure 1 (b) shows that 61% of the sessions achieve a precision = 0.5. Also, 27% of the sessions achieve a coverage = 0.9, while 49% of the sessions achieve a precision = 0.9. On the other hand, after compression of 80% of the URLs (obtained at S_{min} = 0.33), 32% of the sessions achieve a coverage = 0.5, while 42% of the sessions achieve a precision = 0.5. Also, 16% of the sessions achieve a coverage = 0.9, while 33% of the sessions achieve a precision = 0.9. Therefore, compression does affect the quality of the mined profiles. The validation method tries to quantify the trade off between data compression and the quality of mined profiles. Note that in addition to computational and space

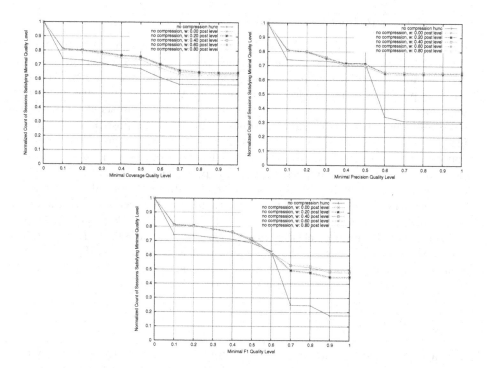

Fig. 3. Library data: Effect of different robustness levels used during post-processing on the achieved cumulative quality Q (on y-axis) measured in terms of (from left to right): coverage, precision, F_1, versus the minimum quality threshold Q_{min} (on x-axis). <u>Legend explanation:</u> *no compression hunc:* results without post-processing, all others are with post-processing at varying robustness levels, W: Minimum robust weight threshold (W_{min}) used in post-processing *robust* profiles (W:0.00 means that no noise is excluded while post-processing, i.e. not robust).

savings, URL compression alleviates data sparsity. Hence it may help uncover smaller profiles, particularly in large and very sparse data.

Figure 1 shows that as expected, compression decreases the final quality of the raw profiles (i.e. no post-processing). However the difference in quality is wider (up to 30% decline) at medium ranges of quality thresholds on coverage and precision (0.1 to 0.5). For higher ranges, the difference is attenuated to the 10% decline range. The legend in each figure lists as (c : and r : respectively) the similarity threshold S_{min} and resulting compression ratio equal to $100(N_U - N_U')/N_U$. Obviously as S_{min} is decreased, the compression ratio increases. However the effect of the *strength* of the compression is not significant on the final quality, as all various compression levels differ by less than 2% in coverage quality, but the difference becomes more acute in the case of precision, where it can reach up to 10%. That said, we notice that based on the overall F_1 measure, all compression levels perform at similar levels, and that final quality is reduced by the effect of compression mostly in mid-range F_1 levels (0.1 - 0.5).

5.3 Effect of Robustness to Noise

Figure 2 shows the effect, on the same session data as above, of post-processing the user profiles by averaging the frequencies, as well as the effect of increasing the minimum robust weight threshold (W_{min}) on the final quality. We can see that post-processing results in striking improvement in coverage, however precision is decreased. We also notice that a higher W_{min} threshold (and hence more robustness to noise) tends to reduce coverage of the post-processed profiles slightly, but generally leads to a significant increase in precision. This is expected from our previous analysis of the effect of robustness. That said, from the point of view of the combined F_1 measure, post-processing and varying levels of robustness yield similar results for the small data, with an optimal robustness level achieved at $W_{min} = 0.6$. However, this result contrasts with the one shown in Figure 3 for a bigger data set collected from the Web server logs of a main university library website, with 11,036 sessions and 8,656 URLs. In this case, post-processing with a robustness level that is not too high ($W_{min} < 0.8$), results in significantly improved coverage, precision, and F_1 measures. For both data sets, the optimal value of robustness level is $W_{min} = 0.6$.

6 Conclusion

We have examined the incorporation of simple cues from a website hierarchy in order to relate clickstream events that would otherwise seem unrelated, and studied their effect on data reduction and on the quality of the resulting knowledge discovery. We have tried to quantify the effect of varying the URL compression strength on the final quality of the user profiles by first merging compatible URLs based on varying thresholds in their URL similarities, and then performing clustering to discover user profiles. Our theoretical analysis showed that significant URL compression can occur, and that this compression will result in a decreased sparsity of the transaction data, which therefore may help the clustering process. We notice that based on the overall F_1 measure, all compression levels performed at similar levels, and that final quality is reduced by the effect of compression mostly in mid-range F_1 levels ($0.1 - 0.5$), but is not affected in the high quality ranges (> 0.8).

Web usage data is also notorious for containing high amounts of noise, thus motivating the use of robust knowledge discovery algorithms that can resist noise and outliers with varying degrees of resistance or robustness. Therefore, we have also examined the effect of *robustness* on the final quality of the knowledge discovery by studying the effect of varying the *robustness level* on the quality of the user profiles. We have also studied the effect of post-processing the user profiles. In the case of a data set with 1704 sessions and 343 URLs, post-processing and robust profiles resulted in higher coverage and lower precision, with the F_1 measure most optimal for robust profiles with a level of robustness $W_{min} = 0.6$. However, in the case of a (10 times) bigger data set (with roughly 30 times more URLs than the first data set), post-processing with a robustness level that is not too high ($W_{min} < 0.8$), results in significantly improved coverage, precision, and

F_1 measures. Too much robustness will exclude too many (including even good) sessions from the profile computation.

To conclude, deciding on *whether to compress the URLs* based on an implicit website hierarchy may depend on the *urgency of the need to significantly reduce the size of the data* for the purpose of *faster* data mining. Deciding on *whether to post-process the discovered user profiles*, and *whether to transform them into robust profiles* would depend on *the final goals* and uses for which the mined profiles are intended, and would take into consideration the *relative importance* of precision and coverage of the robust profiles, and their importance for the website application at hand. For example, if the profiles are to be used as part of a recommendation scheme, then recall and precision are two important criteria that must be weighed carefully in light of the particular domain.

Acknowledgments

We are grateful to the anonymous reviewers for their constructive and helpful comments. This research was supported by National Science Foundation CA-REER Award IIS-0133948 to O. Nasraoui.

References

1. Karnam, P., Joshi, A., Punyapu, C.: Personalization and asynchronicity to support mobile web access. In: Workshop on Web Information and Data Management, ACM 7th Intl. Conf. on Information and Knowledge Management. ACM Press, New York (1998)
2. Agrawal, R., Srikant, R.: Mining generalized association rules. In: 21st VLDB Conference, Zurich (September 1995)
3. Berendt, B.: Understanding web usage at different levels of abstraction: coarsening and visualizing sequences. In: Kohavi, R., Masand, B., Spiliopoulou, M., Srivastava, J. (eds.) WEBKDD 2001 - Mining Web Log Data Across All Customers Touch Points. LNCS (LNAI), vol. 2356. Springer, Heidelberg (2002)
4. Borges, J., Levene, M.: Data mining of user navigation patterns. In: Abbass, H.A., Sarker, R.A., Newton, C.S. (eds.) Web Usage Analysis and User Profiling. LNCS, pp. 92–111. Springer, Heidelberg (1999)
5. Buchner, A., Mulvenna, M.D.: Discovering internet marketing intelligence through online analytical web usage mining. SIGMOD Record 4(27) (1999)
6. Chakrabarti, S., Dom, B., Agrawal, R., Raghavan, P.: Using taxonomy, discriminants, and signatures for navigation in text databases. In: 23rd VLDB Conference, Athens, Greece (1997)
7. Cooley, R., Mobasher, B., Srivastava, J.: Web mining: Information and pattern discovery on the world wide web. In: IEEE Intl. Conf. Tools with AI, Newport Beach, CA, pp. 558–567. IEEE Computer Society Press, Los Alamitos (1997)
8. Cooley, R., Mobasher, B., Srivastava, J.: Data preparation for mining world wide web browsing patterns. Journal of knowledge and information systems 1(1) (1999)
9. Dai, H., Mobasher, B.: Using ontologies to discover domain-level web usage profiles. In: 2nd Semantic Web Mining Workshop at ECML/PKDD-2002 (2002)

10. Eirinaki, M., Lampos, H., Vazirgiannis, M., Varlamis, I.: Sewep: Using site semantics and a taxonomy to enhance the web personalization process. In: ACM conference on Knowledge Discovery in Data, Washington DC, USA (August 2003)
11. Ganesan, P., Garcia-Molina, H., Widom, J.: Exploiting hierarchical domain structure to compute similarity. ACM Trans. Inf. Syst. 21(1), 64–93 (2003)
12. Halkidi, M., Batistakis, Y., Vazirgiannis, M.: On clustering validation techniques. Journal of Intelligent Information Systems 17(2–3), 107–145 (2001)
13. Holland, J.H.: Adaptation in natural and artificial systems. MIT Press, Cambridge (1975)
14. Joshi, A., Weerawarana, S., Houstis, E.: On disconnected browsing of distributed information. In: Seventh IEEE Intl. Workshop on Research Issues in Data Engineering (RIDE), pp. 101–108. IEEE Computer Society Press, Los Alamitos (1997)
15. Klir, G.J., Yuan, B.: Fuzzy Sets and Fuzzy Logic. Prentice-Hall, Englewood Cliffs (1995)
16. Levene, M., Borges, J., Loizou, G.: Zipf's law for web surfers. Knowl. Inf. Syst. 3(1), 120–129 (2001)
17. Mladenic, D.: Text learning and related intelligent agents. IEEE Expert (July 1999)
18. Mobasher, B., Dai, H., Luo, T., Nakagawa, M.: Effective personalizaton based on association rule discovery from web usage data. In: ACM Workshop on Web information and data management, Atlanta, GA (November 2001)
19. Nasraoui, O., Goswami, S.: Mining and validating localized frequent itemsets with dynamic tolerance. In: SIAM conference on Data Mining, Bethesda, MD, USA (April 2006)
20. Nasraoui, O., Krishnapuram, R.: A new evolutionary approach to web usage and context sensitive associations mining. International Journal on Computational Intelligence and Applications - Special Issue on Internet Intelligent Systems 2(3), 339–348
21. Nasraoui, O., Krishnapuram, R.: A novel approach to unsupervised robust clustering using genetic niching. In: IEEE International Conference on Fuzzy Systems, New Orleans, pp. 170–175. IEEE Computer Society Press, Los Alamitos (2000)
22. Nasraoui, O., Krishnapuram, R.: One step evolutionary mining of context sensitive associations and web navigation patterns. In: SIAM conference on Data Mining, Arlington, VA, pp. 531–547 (2002)
23. Nasraoui, O., Krishnapuram, R., Frigui, H., Joshi, A.: Extracting web user profiles using relational competitive fuzzy clustering. International Journal of Artificial Intelligence Tools 9(4), 509–526 (2000)
24. Nasraoui, O., Krishnapuram, R., Joshi, A.: Mining web access logs using a relational clustering algorithm based on a robust estimator. In: 8th International World Wide Web Conference, Toronto, Canada, pp. 40–41 (1999)
25. Nasraoui, O., Krishnapuram, R., Joshi, A.: Mining web access logs using a relational clustering algorithm based on a robust estimator. In: NAFIPS Conference, New York, NY, pp. 705–709 (June 1999)
26. Nasraoui, O., Soliman, M., Badia, A.: Mining evolving user profiles and more: A real-life case study. In: Data Mining meets Marketing workshop, New York, NY, USA (November 2005)
27. Oberle, D., Berendt, B., Hotho, A., Gonzalez, J.: Conceptual user tracking. In: Atlantic Web Intelligence Conference (AWIC), Madrid, Spain (2003)
28. Perkowitz, M., Etzioni, O.: Adaptive web sites: an ai challenge. In: Intl. Joint Conf. on AI (1997)
29. Perkowitz, M., Etzioni, O.: Adaptive web sites: Automatically synthesizing web pages. In: AAAI 98 (1998)

30. Shahabi, C., Zarkesh, A.M., Abidi, J., Shah, V.: Knowledge discovery from users web-page navigation. In: Proceedings of workshop on research issues in Data engineering, Birmingham, England (1997)
31. Spiliopoulou, M., Faulstich, L.C.: Wum: A web utilization miner. In: Proceedings of EDBT workshop WebDB98, Valencia, Spain (1999)
32. Srivastava, J., Cooley, R., Deshpande, M., Tan, P.N.: Web usage mining: Discovery and applications of usage patterns from web data. SIGKDD Explorations 1(2), 1–12 (2000)
33. Terveen, L., Hill, W., Amento, B.: Phoaks - a system for sharing recommendations. Comm. ACM 40(3) (1997)
34. Yan, T., Jacobsen, M., Garcia-Molina, H., Dayal, U.: From user access patterns to dynamic hypertext linking. In: Proceedings of the 5th International World Wide Web conference, Paris, France (1996)
35. Zaiane, O., Han, J.: Webml: Querying the world-wide web for resources and knowledge. In: Workshop on Web Information and Data Management, 7th Intl. Conf. on Information and Knowledge Management (1998)
36. Zaiane, O., Xin, M., Han, J.: Discovering web access patterns and trends by applying olap and data mining technology on web logs. In: Advances in Digital Libraries, Santa Barbara, CA, pp. 19–29 (1998)

From World-Wide-Web Mining to Worldwide Webmining: Understanding People's Diversity for Effective Knowledge Discovery

Bettina Berendt and Anett Kralisch

Institute of Information Systems, Humboldt University Berlin
http://www.wiwi.hu-berlin.de/{~berendt|~kralisch}

Abstract. Users are well-established objects of analysis in Web mining: Web usage mining investigates users' behaviour, Web content and structure mining analyze the content and link structures they generate, Web community mining transfers these questions from analyses of individuals to analyses of groups, etc. However, too often users are reduced to the digital data they have created and/or accessed, and it is (generally implicitly) assumed that "all users are alike" in the ways in which they create and access those data. We argue that to make these analyses and findings more meaningful, a shift is needed from technology to human aspects. This shift calls for a multidisciplinary approach that integrates insights from behavioural, psychological, and linguistic sciences into the field of knowledge discovery. In this paper, we introduce the concept *ubiquity of people* to emphasize that data and knowledge are created and accessed globally, from users who differ in language, culture, and other factors. The Web is the major medium for these activities. The paper investigates how knowledge discovery, including but not limited to Web mining, may benefit from an integration of the concept of ubiquity of people. We provide an overview of the impact of language and culture on how data and knowledge are accessed, shared, and evaluated. We describe a series of studies as an example of integrating these questions into Web (usage) mining. We conclude with a discussion of research questions that are raised by the integration of the ubiquity of people into knowledge discovery, in particular with regard to data collection, data processing, and data presentation.

Keywords: User-centred knowledge discovery, Web usage mining, ubiquitous knowledge discovery, culture, multilingualism.

1 Introduction: Human Factors are Important for Web Mining

Knowledge discovery is more than the application of algorithms – it encompasses the whole process of turning data into knowledge: business / application understanding, data understanding, data preparation, modelling, evaluation, and deployment. People play a pivotal role in this process: people create data, data and knowledge are about people, and people are (or some of them should be) the ultimate beneficiaries of the discovered knowledge. Data-creating activities include the authoring of documents and of references and links between documents, explicit reactions to questions such as

B. Berendt et al. (Eds.): WebMine 2006, LNAI 4737, pp. 102–121, 2007.

the input of registration data, and the behaviour that leaves traces, biometric measurements, etc. in log files.

Because of this central role of people, an understanding of users is required for application / business understanding, for data understanding, for the evaluation of discovered patterns, for the deployment of results, and for all other KD activities that depend on these steps. Because non-users are often affected by data mining, "understanding users" should ideally be replaced by "understanding users and other stakeholders" (Gürses, Berendt, & Santen, 2006). However, in accordance with the dominance of *user* studies in the literature surveyed here, most of this article will restrict its attention to users.

Understanding users involves understanding differences between users. While economic differences and some psychological differences (such as learning styles) have been recognised as a factor in KD and HCI for some time, most studies investigate groups of users living in one country and speaking one language, and they may find differences within these groups. This can become a problem both for understanding and catering to worldwide users. The purpose of the present article is to give an overview of empirical studies on the effects of differences between worldwide groups on behaviours and attitudes relevant to IT usage. We focus mainly on user behaviour regarding the Web since it is today the primary example of a global information source. The purpose of the overview is to argue that integrating these insights into KD is a key step from "mining the World Wide Web" to "worldwide (web)mining", meaning KD processes that integrate knowledge about differences between users.

The contributions of the present article are threefold: First, we introduce the concept of the ubiquity of people, and second, we propose a first general framework for integrating this concept into KD. Third, we illustrate the use of the concept for Web mining using a series of three Web usage mining studies that considered cultural, linguistic and domain expertise factors in a case study

After introducing the new concept of ubiquity of people in Section 2, we argue in Section 3 how KD and its applications can benefit from insights about it. In Section 4, language and culture are investigated as prime aspects of the ubiquity of people, and other aspects are mentioned. The case study is described in Section 5. Implications for KD are outlined in the concluding Section 6.

2   Beyond a Technologically Ubiquitous Web: Ubiquity of People as a New Concept for User-Centred Knowledge Discovery

"Ubiquity" is the property or ability to be present everywhere or at several places at the same time (Oxford English Dictionary). "Ubiquitous" is one of the most characteristic traits of the Internet, referring to its (potential) global access and use. The W3C consortium defines "ubiquitous web" as "web access for anyone, anywhere, anytime, using any device" (http://www.w3.org/UbiWeb/). Other common uses of "ubiquitous" in computing contexts emphasize characteristics of devices, data, and processing (embedded, mobile, spatial/temporal, distributed). These notions focus on technological aspects and only marginally interpret the word in geographical terms.

However, with increasing technological progress, human factors come into the foreground, replacing the erstwhile importance of technological challenges. With a shift from technological areas towards human aspects, the focus in ubiquity research shifts in equal measure to the ubiquity of people.

Data and knowledge sources such as the Internet are accessed and used by users that grew up and reside in different parts of the world. *Ubiquity of people* consequently encompasses diversities with regard to language and culture, but also divergences in economic and social status, technological skills and educational skills. Often, these factors are intertwined and cannot be clearly separated. A connection between language and technological issues is for example given by the fact that languages with a writing system based on Roman characters (e.g., English, German, French) face significantly fewer technical problems in publishing information than languages with different writing systems (e.g., Mandarin, Inuktitut).

Why do we introduce a new concept, and what are its relations to other meanings of "ubiquitous"? The concept "ubiquity of people" can serve as one bridge from technology ubiquity to the goal "global (equal) access". (It is only one bridge because technical, economical, political, etc. factors also play important roles.) Operationally, the concept calls for extensions to user and context modelling, and for extensions to system design. To what extent these extensions only concern content (e.g., further attributes in user models, different layout options) or also formal aspects of modelling and implementation details, is a question for further research. We will discuss both the "why" and "how" of the concept in more detail in the following section.

We focus on linguistic and cultural differences since diversity with regard to these aspects is the most apparent and important in a global context. Global access to data and knowledge eliminates the constraints that were long imposed by geographical location. At the same time, globalisation poses the challenge of meeting individuals' divergent abilities, perceptions, and preferences.

3 How Does KD Benefit from Insights About Ubiquity of People?

The increasing speed of technological progress makes it possible to regard technological barriers as obstacles that can usually be overcome within short periods of time. In contrast, diversities that are inherent to the users, such as their linguistic and cultural backgrounds, are much more stable over time and should be regarded as an established fact for which alternative solutions need to be found.

Such heterogeneity among users leads to heterogeneity in data sets. Data sets are also heterogeneous because data is generated in different contexts. For example, if the market share of a service within a group of native speakers is used as an indicator of a service's success, a comparison between different services and different groups of native speakers is more accurate if the number of potential alternative services is considered as well. Hence, in order to extract meaningful information and knowledge from a global data set, the diversity of contexts needs to be taken into account. To recognize patterns, ubiquitous data processing must incorporate human factors of accessing and using data and knowledge.

The integration of human aspects into knowledge discovery calls for an interdisciplinary approach. Psychological, behavioural, and linguistic sciences provide insight into divergences between users and potential barriers for knowledge acquisition and generation. These disciplines investigate which variables may determine whether or not and to which extent people are able to access data and knowledge, how they evaluate information, and to which degree they are willing and able to share information.

This type of background knowledge is a prerequisite for efficient and accurate data collection as well as for meaningful and correct data processing and interpretation. To give an example: as shown later in this article, an individual's cultural background is a major determinant of his/her attitude towards privacy and data disclosure which in turn might affect the ease of data gathering and the correctness of the data provided.

Finally, one can derive guidelines on appropriate data and knowledge presentation: either directly from insight about the impact of the users' cultural and linguistic background or indirectly from results obtained from data processing that takes cultural and linguistic factors into account. This raises the question which design choices diminish the "digital divide", i.e. which design choices help to provide equal access and encourage participation in knowledge acquisition and generation. Design choices regard technologies that are able to bridge the gap between linguistically and culturally divergent users as well as technologies that are adapted to the different needs of different users.

Thus, a focus on the ubiquity of people contributes to equal *access* to data and knowledge. An accommodation of the ubiquity of people can be understood as the antonym of digital divide: it aims to assure that people independently of their locations, linguistic, cultural, or social backgrounds are able to use the Internet or other global services as information sources. In a second step, this also includes a successful knowledge exchange and knowledge generation across linguistic and cultural borders. This is, for example, particularly important for international work groups or distance-learning groups.

Why not just find out that (say) German Web users like layout type X and US Web users like layout type Y, provide two versions of the software, and regard this as sufficient for globalisation/localisation? While simple solutions like this one may be applicable in certain circumstances, our examples below illustrate that the empirical findings often call for a more differentiated view. We will argue that the ubiquity of people consists of a number of (relative stable) "traits" and (relatively dynamic) "states" that often arise from the interaction of a user's traits with the environmental context. In user and context modelling (cf. Heckmann, 2005; Jameson, 2001), traits and states are often assembled in the user model (e.g., demographics, personality variables, skills, emotional and physical states), properties of the environment are assembled in the context model (e.g., weather conditions), and interactions are modelled as influence relations. The concepts we investigate show the importance of dynamics and of the interaction between user and context. For example, a person's "culture" is often equated with her country of origin, but it may also be the country where she has spent most of their recent life, and it may extend to, e.g., the professional culture a person inhabits. Both may shift over time. Similarly "linguistic background" generally refers to a person's native language. On the other hand, whether a person operates in her native language or in a non-native language is a function also of the environment (here, the language of the information accessed). The same holds for domain knowledge.[1]

4 User Diversity: The Role of Language and Culture

In this section, we provide an overview about cultural and linguistic studies that analyse the impact of culture and language. Their results provide a first outline of background knowledge necessary for successful ubiquitous data collection and processing. We emphasise how culture and language affect (1) access to data and knowledge, (2) people's willingness and ability to share it, and (3) their evaluation of information. These three aspects interact in several ways, so a clear line cannot always be drawn. If, for example, access to data and knowledge is difficult or impossible, sharing of information is restricted as well.

4.1 Language

When a user accesses a service, she may access content presented in her first (native) language and thus find herself in an "L1 situation". She may also access content in a second (non-native) language and thus find herself in an "L2 situation". Since this distinction is the most basic and best-researched variation in language, we will operationalize "language" as L1 vs. L2.

Access to Data and Knowledge. Availability of content differs between languages to a major extent. A myriad of articles (e.g. Danet & Herring, 2007) describe the original dominance of English language content on the Internet. Over the years the mono-lingual dominance has been increasingly counterbalanced by other, widely spoken languages (see statistics by Internet World Stats, 2007). Nevertheless, the majority of languages is still underrepresented on the Internet and will probably never attain an adequate and proportional representation. This can be attributed to various factors such as the number of native speakers, their economic importance, or the distribution of the Internet in certain areas. These factors diminish the importance of a group of native speakers as a target group and limit at the same time the number of potential native language website creators.

If native language (L1) content is not available or limited to a few topics, users are forced to access information in a non-native language (L2).[2] The L2 is usually English, and sometimes the area's lingua franca, such as Russian for communication within the former Russian republics (Wei & Kolko, 2005). Depending on a user's L2 proficiency levels, access to data and knowledge might not be possible or reduced to a

---

[1] An interesting extension would be a differentiation between "acting in a familiar culture" and "acting in an unfamiliar culture".

[2] The shift to an L2 situation may also involve other causes, including the possibility to access a larger repository, the wish to compare different opinions, the need to retrieve information in English because it is better exploitable in other contexts.

minimum amount. Palfreyman and Al-Khalil (2003) find that even in a diglossic[3] situation, such as commonly found in Arab countries, the use of the high dichotomy (= standard language) often constitutes a major barrier for users with lower education.

Kralisch and Mandl (2006) and Halavais (2000) show by means of logfile analyses that websites are indeed favoured by native speakers, even if the respective percentages of native speakers and content alternatives are taken into account.

The ease of reading information is only one aspect of the accessibility of data and knowledge. In fact, accessibility is also determined by the ease of rendering available information. From a technological point of view, restrictions in the usage of characters present a significant inconvenience for certain language groups. The ASCII Code, originally based on the English language, favours writing systems that are derived from the Latin alphabet ("typographical imperialism" – Phillipson, 1992; Phillipson & Skutnabb-Kangas, 2001). Speakers of languages that are based on different writing systems (e.g. Cyrillic alphabet, Chinese signs) are disadvantaged by the ASCII code (Pargman & Palme, 2004) and forced to find work-arounds such as visual numbers (Palfreyman & Al-Khalil, 2003; Tseliga, 2007). Also, due to the still common use of ASCII code, access to less widespread writing systems (e.g Inuktitut) usually requires downloading the specific character set (Herring & Estrada, 2004). The introduction of Unicode, which covers almost all writing systems in current use today, is an essential step towards multilingual content generation and multilingual computer processing and hence towards equal accessibility.

Sharing Data and Knowledge. Language may constitute a barrier for sharing data and knowledge. With increasing distribution of the Internet, the role of the English language as the Internet's lingua franca is reduced: the percentage of English native speakers and highly proficient L2 users decreases, whereas the number of different native languages grows (Internet World Stats, 2007). Consequently, communication barriers due to the lack of a common language increasingly arise.

Herring et al. (2006) discuss, in a study of language networks on LiveJournal, the role of language as a determinant of network generations and in consequence as a determinant of data and knowledge exchange which occurs mainly *within* each network. These networks' sizes and densities differ between languages. Herring et al. explain these divergences by differences in the numbers of users per language and in the degree of bilingualism and point out that a critical mass is necessary to create a robust language network. Similar phenomena are reported about the multilingual European discussion forum Futurum. Although participation in the language of choice is encouraged (a battery of translators ensure translation between languages), communication in English clearly dominates. Moreover, discussion threads that are introduced in other languages tend to be shorter (Wodak & Wright, 2004).

Since multilingualism will increase rather than diminish with the growing distribution of global technology[4], data gathering and data processing will more and more be required to take these barriers into consideration as well as consider the use of multilingual technologies.

---

[3] In simplified terms, diglossia describes a linguistic situation in which a standard form of a language with high prestige (e.g. classic Arabic) and a dialect form with lower prestige (e.g. the regional forms of Arabic) are spoken in a society. For details, see (Ferguson, 1959).

[4] This expectation is based on past and current developments, see (Global Reach, 2006).

Evaluation of Information. In contrast to culture (see below), the impact of language on information evaluation is rather indirect. Language predominantly determines how easy it is to access data and knowledge. According to Davis' (1989) model of technology acceptance (TAM), ease of use is a determinant of usefulness, attitude towards an information system, and satisfaction. It can therefore be assumed that information in a user's native language leads to a more positive evaluation, by that user, of the information (Kralisch & Köppen, 2005). This effect might be strengthened if the native language is perceived as an identity-constituting factor.

4.2 Culture

"Culture" is a multifaceted and controversial term. In its most general meaning , it denotes attitudes and behaviours of a group that are relatively stable over time, and the term is also used to denote the group united by these commonalities. Many studies that are relevant for IT understanding and design have operationalized culture as a country or a collection of countries. While we are aware of the problems induced by this reading of "culture", we adopt it as a working definition (for a detailed discussion, see Kralisch, 2006). The reasons are the predominance in the literature and the relevance for applications that often aim at understanding and opening a new market that is defined by the same boundaries: countries or country groups.

Access to Data and Knowledge. In contrast to language, culture represents a less visible obstacle to data and knowledge access. However, the Internet itself "[...] is not a culturally neutral or value-free space in which culturally diverse individuals communicate with equal ease" (Reeder, Macfadyen, Chase, & Roche, 2004). It is a cultural product that reflects the cultural values of its Anglo-American producers. "Their ... cultures value aggressive/competitive individualistic behaviours. ... These cultural value communications are characterized by speech, reach, openness, quick response, questions/debate and informality" (Reeder et al., 2004). A number of studies investigate the impact of the Internet's cultural values on accessing data and knowledge. Their results provide a differentiated picture with sometimes contradictory outcomes.

Reeder et al. (2004) state that due to the implicit cultural values embodied by the Internet, English speaking, academic cultures have the least difficulty in communicating over the Internet (see also De Mooij, 2000).

Hongladarom (2004) describes the efforts carried out by the Thai government to encourage Internet access among all classes of Thai society. Despite the government's technological and economic efforts (such as hardware and software support), the success of this initiative is rather limited. Hongladarom considers this to be a result of the Internet's implicit cultural values which contradict traditional Thai values. These results are in line with Warschauer's (2003) argumentation that the digital divide is interrelated with socio-cultural factors. Hongladarom's and Warschauer's findings are contradicted by Dyson's study (2004). Dyson finds in his analysis of adoption of Internet communication technologies among indigenous Australians that the cultural values of the Internet do not represent an obstacle for accessing data and knowledge. Other factors are pointed out by the author as causes of the low adoption rate.

Beside the negative or neutral impact of the Internet's cultural values, the characteristics of the Internet have also been shown to positively affect the participation of users from societies with divergent cultural values. Several studies point out that the impersonality of Internet communication encourages women from traditional, high Power Distance countries (e.g. Thailand) to participate more freely in Internet use and communication (Panyametheekul & Herring, 2003; Wheeler, 2001). A similar effect was shown for Asian students. Asian teaching styles are characterized as authoritative with strong hierarchies between students and teachers. The impersonal communication style on the Internet encourages Asian students to participate more during lessons (e.g. Bauer, Chin, & Chang, 2000).

Sharing Data and Knowledge. Whereas language predominantly affects the *capacities* for sharing data and knowledge, culture has a major impact on users' *willingness* to share information. The literature suggests that self-conception, attitude towards privacy, and a cultural groups' hierarchical organisation are the most important factors for the willingness to share information.

A group's self-conception and cultural identity play a major role with regard to the conception of ingroups and outgroups and its impact. The strength of ingroup and outgroup perception is strongly correlated with the cultural dimension of individualism and collectivism[5]: collectivistic cultures – in contrast to individualistic societies – tend to strongly differentiate between ingroup and outgroups. Various studies indicate that, as a consequence, traditional and collectivistic cultures fear fraud and moral damage through the information that is provided on the Internet.

Privacy issues have also been shown to be affected by a culture's degree of individualism. Members of individualistic cultures tend to be less willing to provide sensitive information than members of collectivistic cultures. This can be explained by the observation that individualistic cultures value private space more than collectivistic cultures (e.g. Milberg, Smith, & Burke, 2000).

Individualistic and collectivistic cultures also differ in the type of information they provide when negotiating identity (Burk, 2004; Reeder et al., 2004). "It is likely that in some cultural settings, information considered highly personal by Western standards, such as wealth or spending habits, may be deemed open and public, whereas information considered relatively innocuous in Western settings, such as a nickname, might be considered extremely private" (Burk, 2004). Debnath and Bhal (2004) point out that ethical issues related to privacy differ among Indian citizens depending on their acquired norms of ethical and moral conduct. Burk (2004) emphasizes that "privacy as a matter of individual autonomy may be relatively unimportant in cultural settings where communal information is unlikely to be accommodated within the data protection models now sweeping across the globe".

In addition, power distance[6] has a similar impact on users' willingness to disclose data, with members of high power distant countries being more willing to provide data than members of low power distant countries (Kralisch, 2006). However, within high power distant societies, knowledge sharing from high hierarchy levels to low

---

[5] Individualism and Collectivism are cultural dimensions developed by Hofstede (1991). Individualism implies loose ties between the members of a society; collectivism implies that people are integrated into strong, cohesive groups (Marcus & West Gould, 2000).

[6] One of Hofstede's cultural dimensions that describes the extent to which less powerful members of institutions and organisations accept that power is distributed unequally (Hofstede, 1991).

hierarchy levels is difficult since it would transfer decision making authorities to subordinates. Heier and Borgman (2002) describe how this effect challenged the international HRbase, an Intranet-based knowledge management system, of Deutsche Bank: usage rates were about 20% in Germany and the UK but only about 4% in Asian countries.

Evaluation of Information. Differences in information evaluation are strongly related to differences / compliances in communication styles. In particular, the preference for face-to-face (personal) communication over impersonal communication is pointed out as an important cultural factor of Internet use and information evaluation. High preference for personal communication usually leads to negative evaluation of information from outsider groups; this information is deemed less reliable. Reliability of information is attributed in these cultures to the reliability of its carrier. Technologies are not seen as an equivalent of interpersonal communication and are therefore not as trustworthy. "Technologies that facilitate interpersonal connections among people who want to stay in touch (e-mail, cell-phones) would [therefore] be adopted much faster than impersonal devices (..., web-info sites, e-commerce, call-centers, automated messages)" (Markova, 2004). In line with these findings, results from a study conducted by Siala and his colleagues (2004) reveal that collective cultures buy mainly from within-group members. Similarly, Jarvenpaa and Tractinsky (1999) found that trust in e-commerce is culturally sensitive.

Furthermore, members of different cultural groups have different approaches towards contradicting information and its evaluation. Again, the level of power distance appears to be important: members of high power distant countries tend to more easily accept information unquestioningly. Markova (2004) describes the cultural concept of information in central Asia: information is not searched or evaluated, but memorized the way it is "taught". The cultural belief in objective truth is supported by government-controlled accessed to information that inhibits access to conflicting information. This finding is in line with the teaching styles in high power distant countries (see "knowledge sharing across hierarchical levels" above).

Finally, culture also shapes assumptions about which knowledge is important (DeLong & Fahey, 2000). De la Cruz et al. (2005) show that members of different cultures assign different importance to the same website criteria. As a consequence the quality of websites is interpreted differently. More detailed research is however required in order to specify the relationship between cultural values and importance of information elements.

Last but not least, depending on their cultural background users differ in their way they express their evaluation. Evers and Day (1999) have therefore developed recommendations that help normalize evaluations from users with different cultural backgrounds. For example, they recommend 6-point Likert scales to avoid neutral positions that are often adopted by members of collectivistic cultures.

4.3   Factors Beyond Language and Culture

We introduced ubiquity of people as a term that describes people's diversity and the call for the provision of equal access. In view of increasing globalisation we focused on linguistic and cultural aspects. Nevertheless, people's diversity covers more than these traits. We mentioned in the Introduction that differences in technological progress, social and economic status as well as levels of education are further factors that affect access to data and knowledge, sharing and evaluation. These factors often mediate or reinforce the impact of culture and language on user behaviour. An individual's level of education (or domain knowledge) appears to be particularly related to his/her linguistic abilities. Economic/social status and technological progress often strengthen the impact of cultural values but also linguistic abilities.

The results of Kralisch and Berendt (2005) indicate that L2 users with high domain knowledge manifest the same preferences in their search for information as L1 users. In contrast, L2 users with low domain knowledge show divergent preferences. In other words, domain knowledge can compensate for language deficiencies. More generally, a user's education plays a major role in areas with lower Internet distribution and large educational divergences within a society. Education affects access to information since usually only elites have access to the Internet and possess the necessary computer skills (e.g. Mafu, 2004) and/or the literacy to take full advantage of complex content. As described above, in situations of diglossia, higher education usually assures a higher language proficiency level in the standard language, in particular when writing.

Linguistically and culturally determined difficulties in accessing data and knowledge are often complicated by lack of access to technologies in remote areas or high access fees. Dyson (2004), for example, attributes the lower adoption rate among indigenous Australians to limited access to Internet communication technologies, to high costs, poor telecommunication infrastructure, and low computer skills.

5 Ubiquity of People as a Factor in Web Mining: Examples

So how can findings like those described in Section 4 be used in Web mining? In this section, we give an overview of a series of studies from our own research. In these studies, findings on the effects of people's diversity were employed as an independent variable or as background knowledge.

5.1 Data and Data Preparation

Data were taken from the logfile of a large and heavily-frequented international website, recorded between November 2001 and November 2002. The site is a public health-care information site, and it was available in four languages (English, German, Spanish, and Portuguese) at the time of data collection. All language versions are presented with the same interface design.

Sessionizing, robot elimination, and basic data extraction from the logfile followed standard procedures (e.g., Cooley et al., 1999), using the tool WUMPREP (www. hypknowsys.de). As elsewhere in Web usage mining, sessions were treated as (pseudo-)users; intra-individual maturation effects could therefore not be measured.

Geographic information was obtained from IP addresses using Geoselect (www. geobytes.com). Based on the detailed geographic data, the cultural indices (Hofstede, 1991, Hall, 1989) were assigned to each session. The classification of countries as either monochronic or polychronic follows the demographic scale suggested by Morden

(1999); values range from 0 to 100. IP addresses that did not allow the reconstruction of a user's country were removed. Studies 1 and 2 used data from 55 countries with 5,136 sessions and 54,074 page requests in total. For study 3, a one-country subsample was used. It consisted of 20,333 requests in 1397 sessions.

The data were semantically enhanced by mapping all URLs into a *concept hierarchy* that mirrored (a) the general architecture of an information site and (b) the domain ontology (here: an internationally standardized classification of diseases described in the site as different "diagnoses"). Search-page requests were mapped to the search option employed (alphabetical search, location search – a search criterion based on the content's intrinsic organization, search engine) or to these concepts' super-concept "search". Pages describing a diagnosis in textual, pictorial, or other form were mapped to the ID of that diagnosis or to these concepts' super-concept "diagnosis". All entry pages were mapped to the concept "index page". Other concepts did not play a major role in the identified patterns.

Structurally, sessions were modelled as sequences or (multi)graphs in one of three ways. For example, a session with requests [A,B,C,A,B,D,B] was modelled as that sequence (study 2), as {{A,B,C,D},{(A,B), (B,C), (C,A), (B,D), (D,B)} (directed graph: study 1) or as {{A,B,C,D},{(A,B), (B,C), (C,A), (B,D)} (undirected graph: study 3). Self-loops were eliminated, such that sequences of (different or repeated) accesses to URLs all instantiating the same concept all mapped to one node in the occurrence-numbered sequence or graph.

5.2 Linear vs. Non-linear Navigation Patterns

In (Kralisch, Berendt, & Eisend, 2005; Berendt & Kralisch, 2005), we derived three hypotheses on navigation behaviour from the definitions and previous findings on three prominent cultural indices (see Section 4). Each hypothesis predicted a positive (or negative) effect of the value of an index on the value of a behavioural variable. Two behavioural variables were amount of information gathered and time spent in the site, measured by the number of pages visited and the total dwell-time. The values of these variables could be obtained in a straightforward way from database queries.

The third behavioural variable was more interesting from the viewpoint of Web usage mining: the linearity of navigation patterns. We derived the hypothesis that members of monochronic cultures are more likely to show linear navigation patterns whereas members of polychronic cultures are more likely to show non-linear navigation patterns.[7]

The linearity (or not) of navigation has been investigated for a long time because the essence of hypertext is that it affords non-linear navigation – and the question is to what extent users utilize this opportunity. Characterizations of linear vs. non-linear local patterns have been described in a number of studies since Canter, Rivers, and Storrs (1985). Essentially, linearity is the absence of loops (van Dyke Parunak, 1989). However, there are different ways of measuring the degree of linearity.

---

[7] Hall's (1989) extension of Hofstede's concept of time dimension refers to cultures' approaches towards structuring time. Monochronic (sequential) cultures are characterized by the isolation of activities. People tend to do one thing at a time. The opposite approach towards structuring time is synchronism (polychronic cultures). People structuring time synchronically usually do several things at a time, and plans are easily changed.

5.2.1 Global, (Purely) Structural Linearity

Patterns and measures. A well-known global measure of a graph's linearity is *stratum* (e.g., McEneaney, 2001). It is defined as

$$stratum(G) = \Sigma_{v \in V} \; [\; abs(\Sigma_{u \in \; superordinates(v)} \; l(u,v) - \Sigma_{u \in \; subordinates(v)} \; l(v,u) \;] \; / \; N$$

where V is the set of the graph G's nodes, *superordinates(v)* are all nodes from which v can be reached, *subordinates(v)* are all nodes that can be reached from v, and $l(u,v)$ is the length of the shortest path from u to v. N is a normalization factor to ensure that *stratum* $\in [0;1]$, where 0 (1) describes a completely non-linear (linear) graph.

Since countries were ordered from low values for monochronic to high values for polychronic cultures, we expect a positive correlation between this scale and the linearity measure stratum.

We measured the average stratum per country (averaged over all sessions from that country in the dataset) by transforming each session into a directed graph and computing the graph index stratum, and we examined the correlation with mono-/polychronicity values. The results were inconclusive.

One possible reason for the inconclusiveness is semantics: All nodes contribute to stratum in the same way; however, a (non-)linearity in search pages is a different statement about a user's navigation than the structurally identical (non-)linearity in content pages.[8] Therefore, we defined a new measure.

5.2.2 Local, Semantic Linearity

Patterns and measures. The new measure (Kralisch, Berendt, & Eisend, 2005) was designed to measure *content-based (non-)linearity* describing access to the site' core content. A sequence of actions is deemed linear if no content areas (individual diagnoses) are re-visited. A linear pattern may also contain intermediate (even repeated) visits to search pages that are generally used only for organization.

Linear patterns were defined as all bindings of concept sequences to a template

> search* diagnosis1 search* diagnosis2
> search* diagnosis1 search* diagnosis2 search* diagnosis3

etc., where search are arbitrary search pages, diagnosis1 etc. are pages on distinct diagnoses, and * is 0 or more instances. (Due to the prior elimination of self-loops, diagnosis1 indicates one or more requests for information on this concept.) We defined L as the support of maximal patterns of this type: the absolute number of occurrences of maximal linear patterns in a set of sessions.

Non-linear patterns are defined as patterns with loops, here: loops at the diagnosis level. Non-linear patterns were defined as all bindings of concept sequences to a template

> search* diagnosis1+ search* diagnosis2+ search* diagnosis1+

etc. We defined NL as the support of maximal patterns of this type.

---

[8] Another reason might be the properties of the aggregation measure "arithmetic average". In future work, we plan to explore different statistics like the median. We thank an anonymous reviewer for pointing this out.

For simplicity, all requests that did not map to a search or diagnosis concept had been filtered out prior to mining (they could also be left in, but would require a much more unwieldy regular expression as template). Examples of patterns are shown in Fig. 1. ("D_" indicates the information catalogue within the site.)

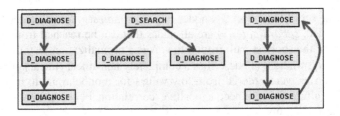

Fig. 1. Examples of linear patterns (left, middle) and of non-linear patterns (right) in study 2

The measure should consider the relative frequency of linear and non-linear patterns in a set of sessions. We define the structural part as follows:

$$local\text{-}linearity\ (\{S\}) = (L - NL)\,/\,(L + NL)$$

This measure has a value range of $[-1; 1]$. The maximum value is reached when there are only linear patterns; the minimum when there are only non-linear patterns.

Mining. To obtain the numbers L and NL, we needed a mining tool that was able to find the support of patterns specified by regular expressions. We used the sequence miner WUM (www.hypknowsys.de). This tool contains a query language that allows the analyst to narrow the search to interesting patterns based on the identities of pages along a path (essentially, in the form of regular expressions like those described above). In addition, constraints on support and confidence can be expressed. WUM can hence be used for various combinations of exploratory and confirmatory analyses. Mining operated on subsamples defined by the country from which its session came.

The minimum support threshold was set to 1 because pattern restriction was by shape and content rather than by support. WUM finds the support and confidence of each binding to a template given by the analyst; from these statistics, the number of paths that instantiate a template can be derived. WUM results were post-processed manually to extract the maximal patterns, such that each request was counted at most once towards a pattern.

Results and interpretation. There was a significant (p < .001) Pearson's product-moment correlation of r=.196 between mono-/polychronicity and linearity/non-linearity. Even if the absolute value of the correlation is not very high, this result corroborated the hypothesis.

We controlled for the effects of native vs. non-native speakers: no effect of native and non-native countries on cultural dimensions under investigation was found.

These findings indicate that structural measures of navigation should be considered relative to the specifics of a site. For example, "linearity in information gathering" can only be meaningfully formalised when the semantics of the requested pages are taken into account.

One weakness of this study was its largely confirmatory nature: Only pre-defined patterns could contribute to the result, and differences could only be found between the sets of sessions pre-defined by their values on the mono-/polychronicity scale. The exploratory element was restricted to finding out which diagnoses (etc.) instantiated the pattern templates. In addition, the findings on search-option choice (see Section 4) indicated that we should "drill down" into the characteristics of behaviour associated with different search options. We therefore modified our session representation and mining technique to provide for a completely exploratory study.

5.3 An Exploratory Study of Navigation Patterns

In this study (Berendt, 2006), we looked for frequent (sub)graph patterns in sessions. The search for patterns was not restricted by content or structure as in the first study, but by minimum support. Thus, it was an exploratory search for typical but not pre-constrained patterns of usage.

Patterns, measures, and mining. We mined for arbitrary frequent subgraphs using the tool fAP-IP (Berendt, 2006), which can operate with concept hierarchies. Thus, a pattern can be found on the coarsened concept level (e.g., the path "diagnosis search diagnosis"), and later the analyst can "drill down" into this pattern to find which diagnoses (e.g., 12345 and 23456, but also 6789 and 7891) instantiated this pattern. Search-page requests were mapped to the search option rather than to the general concept "search". Paths, trees and cyclic graphs can be found; a re-writing in terms of sequence templates as above is not necessary. The only measure defined before mining was minimum support as an indicator of pattern importance (3%). Measures of pattern shape were implicit in a qualitative interpretation of the results (see below).

Results: Basic statistics on frequent patterns. Most patterns involved the same concepts as those in study 2, and only those. We concentrate on reporting these. Chains of diagnoses were the most frequent patterns: a chain of six diagnoses had support 7.2% (5: 9.2%, 4: 13%, 3: 18.9%). In addition, patterns with three or more other diagnoses branching off a "hub" diagnosis as shown at the left of Fig. 2 (support 5.3%) were frequent. Rings also occurred at slightly lower support thresholds (see Fig. 2, second from left). "DOIA" in the figure is an index page.

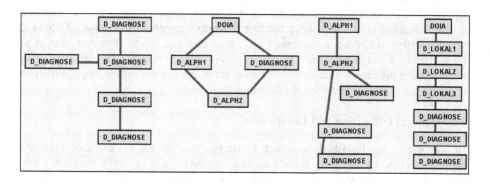

Fig. 2. Frequent navigation patterns in study 3

Results: Search options – linear vs. hub-and-spoke. Patterns illustrated the use of different search options. The internal search engine appeared only in very few patterns (only in 2-node patterns: 4.2% for search-engine and a diagnosis, 3.5% for search-engine and alphabetical search—probably a subsequent switch to the second, more popular search option). This was because the search engine was less popular than the other search options (used about 1/10 as often), but far more efficient in the sense that searches generally ended after the first diagnosis found (assuming that finding a diagnosis was the goal of all search sessions). The alphabetical search option generally prompted a "hub-and-spoke navigation", as shown on the right of Fig. 2 (support 6.4%). In contrast, location search generally proceeded in a linear or depth-first fashion, as shown on the far right of Fig. 2 (support 5%; with one diagnosis less: 6.9%).

Interpretation. This may be interpreted as follows: Location search prompts the user to specify, on a clickable map, the body parts that contain the sought disease. This is in itself a search that can be refined (LOKAL1 – LOKAL2 in the figure; a similar pattern of LOKAL1 – LOKAL2 – LOKAL3, followed by 2 diagnoses, had a support of 5.1%). This narrowing-down of the medical problem by an aspect of its surface symptoms (location on the body) helps the user to identify one approximately correct diagnosis and to find the correct one, or further ones, by retaining the focus on symptoms and finding further diagnoses by following the differential-diagnosis links in the site. Thus, non-expert users in particular can focus on surface features that have meaning in the domain, and they can acquire some medical knowledge in the process.

Alphabetical search, on the other hand, leads to lists of diseases that are not narrowed down by domain constraints, but only by their name starting with the same letter. Navigation choices may be wrong due to a mistaken memory of the disease's name. This requires backtracking to the list-of-diseases page and the choice of a similarly-named diagnosis.

Domain knowledge and navigation behaviour. This interpretation of search options and associated navigation is supported by the findings, reported in Section 4 above, from the study of search-option use in the same site in which participants specified whether they were physicians or patients. Content search was preferred by patients, whereas physicians used alphabetical search or the search engine more often. The linking of the preference for the location-search option on the one hand, and the frequent behavioural pattern associated with content search on the other, showed that a differentiation between distinct forms of "linear" search in the sense of study 2 reveals further differences associated with user characteristics. However, this is at present only a conjecture that, in the available data, could not be validated with demographic information. In future research, these conjectures on language, expertise and navigation should be tested in confirmatory settings.

5.4 General Discussion and Limitations

In summary, the described sequence of studies has shown evidence of different influences of people-ubiquity variables on navigation behaviour, and it has shown that even for a simple construct like "linearity", it is necessary to consider the implications of the operationalizing measure. Future work should investigate the relationships between the different measures in more detail, as well as the interactions between the different user variables language, domain knowledge, and culture.

In addition, methodological challenges of logfile analyses should be addressed. In particular, most personal and situational variables cannot be controlled, a typical limitation of field studies. If systematic biases are assumed, other methods are required in order to investigate those issues, such as additional online questionnaires covering measures on user's attitudes, goals, and strategies. The second major problem is the potential effect of the restriction to one site and domain. This is a common problem in website-usage studies in field and laboratory studies. However, the extension to more than one site and domain is even more difficult for large-scale field studies, since it is very difficult to find a set of real-world, heavily frequented websites that differ only with respect to only one feature. In summary, these studies should be seen in a wider context of research on the ubiquity of people.

6 Conclusions: Research Questions for User-Centred Knowledge Discovery in a Global Context

We presented aspects of how language and culture may influence the way people access data and knowledge, share them, and evaluate them. These aspects were pointed out as necessary background knowledge for user-centred knowledge discovery that deals with the ubiquity of people, and an example from Web usage mining was described.

Research on the impact of user characteristics on interaction with the Web is a lively field of current research with many open questions. For example, the studies presented in Section 4 were mostly intracultural or bicultural comparisons. In order to obtain a wide range of background knowledge, *multi*cultural comparative studies are necessary (see also Danet & Herring, 2003).

We close by discussing research questions in user-centred knowledge discovery that are raised by people's ubiquity. We focus on three aspects of knowledge discovery: data collection, data processing, and data presentation.

Data Collection. Data collection from ubiquitous users must cope with two major problems: challenges of obtaining data and challenges of their representativeness.

Data collection efforts that rely on users' self-reports need to consider that users differ in their ability to provide information as well as in their willingness to share it. Language, for example, may constitute a major barrier to accessing questionnaires or websites and may limit users in their abilities to answers questions. Users' willingness to disclose data is highly culturally determined, as shown by studies of individualistic and collectivistic cultures. A low willingness to share information leads to the question of how reliable the information provided by the user is. Culturally determined differences in privacy issues and willingness to share information ask for a detailed examination of the extent to which data gathering would constitute an intrusion into the private space. As shown above, differences may regard users' general attitudes as well as specific types of information. Privacy research has shown that presenting reasons for collecting data creates confidence among users and augments the amount of data provided by them (Kobsa & Teltzrow, 2005). In a global context, the presented reasons might need to be reviewed and adapted to the local needs and preferences.

Given the differences in accessibility of data and knowledge sources, the question is raised how representative the investigated group is. An English-language Internet questionnaire might for example be answered by a wide range of English native speakers that differ largely in their economic and social status and educational levels. At the same time the questionnaire might cover an only very specific group of Arab native speakers, namely those that are well educated and affluent.

Furthermore, a user's cultural background affects the way opinions are expressed. This should be taken into account through a culturally adapted conception of data gathering tools or through data processing that considers these differences.

Non-reactive methods are challenged by the difficulties of correctly assessing a user's cultural background and linguistic abilities. Analyses of logfiles and IP addresses can be considered only proxies with a limited certainty of the data collected. Analyses of IP addresses are for example used to obtain information about the location from which the Internet is accessed. Information about the location in turn helps to derive information about a user's linguistic and cultural background, but involve a certain error (Kralisch & Berendt, 2005; Kralisch, 2006).

Data Processing. Ubiquity of people leads to heterogeneous data sets due to different contexts. User-centred knowledge discovery hence requires data processing that takes background knowledge about the users and their context into account.

For example, if a user accesses a website despite major linguistic challenges, this might signify a higher relevance of the website's content or service. In cases where a relationship between use/access and relevance (or other attributes) is established, data processing becomes significantly more accurate if weighted measures are used that consider these challenges. In a similar manner, if the amount of generated content/ services is analysed as an indicator of need or interest, context information about difficulties of content/service generation render the analysis more accurate.

Given the increasing amount of multilingual data sets, knowledge discovery should also take into consideration research results regarding multilingual information retrieval tools or information retrieval tools that take cultural aspects into account. For example, Kralisch and Mandl (2005) provide a first overview how the users' cultural backgrounds affect the use of information retrieval tools.

Data Presentation. Information about the impact of language and culture on data and knowledge accessibility provides important insights into suitable forms of data presentation. Further insights can be obtained through appropriate data processing. Culture and language are two factors that affect people's abilities and preferences for certain forms of data presentation. For example, the outcomes of Kralisch and Berendt (2004) indicate that users from high power distant countries prefer a hierarchical form of knowledge presentation more than members of low power distant countries. Kralisch, Berendt, and Eisend (2005) propose design guidelines based on this and related findings. Further divergent preferences were found with regard to other cultural dimensions. In a similar manner, Yeo and Loo (2004) present cultural differences in preferences for classification schemes.

Research on data presentation forms also involves the development of technologies that are able to bridge the gap between different cultures and languages, such as multilingual information retrieval tools. However, where bridging the gap is not conceivable or feasible, adaptations to the users' cultural and/or linguistic needs are necessary. User-centred knowledge discovery should therefore also aim to discover thresholds where adaptations to the user's linguistic and cultural needs are necessary and where other solutions are more efficient and/or appropriate.

In future work, we also plan to investigate the technical implications of these findings. In particular, we intend to explore how the ubiquity of people can be reflected in user and context modelling, and put to use in the processes by which these models enter KD and its deployment for user adaptation (cf. Heckmann, 2005; Jameson, 2001).

References[9]

Bauer, C., Chin, K., Chang, V.: Web-Based Learning: Aspects of Cultural Differences. In: Proc. of the 8th European Conf. on Information Systems (ECIS), July 3-5, Vienna Austria (2000), http://is2.lse.ac.uk/asp/aspecis/20000068.pdf

Berendt, B.: Using and learning semantics in frequent subgraph mining. In: Nasraoui, O., Zaïane, O., Spiliopoulou, M., Mobasher, B., Masand, B., Yu, P.S. (eds.) WebKDD 2005. LNCS (LNAI), vol. 4198, pp. 18–38. Springer, Heidelberg (2006)

Burk, D.: Privacy and Property in the Global Datasphere: International Dominance of Off-the-shelf Models for Information Control. In: Sudweeks & Ess, pp. 363–373 (2004)

Canter, R., Rivers, R., Storrs, G.: Characterizing User Navigation through Complex Data Structures. Behaviour and Information Technology 4(2), 93–102 (1985)

Cooley, R., Mobasher, B., Srivastava, J.: Data preparation for Mining World Wide Web Browsing Patterns. Journal of Knowledge and Information Systems 1(1), 5–32 (1999)

Danet, B., Herring, S.: Introduction: The Multilingual Internet. Journal of Computer-Mediated Communication, 9(1) (2003), http://jcmc.indiana.edu/vol9/issue1/intro.html

Danet, B., Herring, S.: Multilingualism on the Internet. In: Hellinger, M., Pauwels, A. (eds.) Handbook of Language and Communication: Diversity and Change. Mouton de Gruyter, Berlin (2007), http://pluto.mscc.huji.ac.il/~msdanet/papers/multiling.pdf

Davis, F.D.: Perceived Usefulness, Perceived Ease of Use, and User Acceptance of Information Technology. MIS Quarterly 13, 319–340 (1989)

Day, D., Evers, V.: Questionnaire development for multicultural data collection. In: del Galdo, E., Prahbu, G. (eds.) Proc. of the Int. Workshop on International Products and Systems, Rochester, UK (May, 20-22, 1999)

de la Cruz, T., Mandl, T., Womser-Hacker, C.: Cultural Dependency of Quality Perception and Web Page Evaluation Guidelines: Results from a Survey. In: Day, D., Evers, V., del Galdo, E. (eds.) Proceedings of the 7th IWIPS, Amsterdam, The Netherlands, July 7-9, pp. 15–27 (2005)

De Mooij, M.: The Future is Predictable for International Marketers: Converging Incomes Lead to Diverging Consumer Behaviour. International Marketing Review 17(2), 103–113 (2000)

Debnath, N., Bhal, K.: Religious Belief and Pragmatic Ethical Framework as Predictors of Ethical Behavior: An Empirical Study in the Indian Context. In: Sudweeks & Ess, pp. 409–420 (2004)

DeLong, D., Fahey, L.: Diagnosing Cultural Barriers to Knowledge Management. Academy of Management Executive 14(4), 113 (2000)

Dyson, L.: Cultural Issues in the Adoption of Information and Communication Technologies by Indigenous Australians. In: Sudweeks & Ess, pp. 58–71 (2004)

Ferguson, C.: Diglossia. Word 15(2), 325–340 (1959)

---

[9] All online sources were last accessed on June 1, 2007.

Gürses, S.F., Berendt, B., Santen, Th.: Multilateral security requirements analysis for preserving privacy in ubiquitous environments. In: Proc. Workshop on Ubiquitous Knowledge Discovery for Users at ECML/PKDD 2006, pp. 51–64 (2006), //vasarely. wiwi.hu-berlin.de/UKDU06/

Halavais, A.: National Borders on the World Wide Web. New Media and Society. Proceedings/UKDU06-proceedings.pdf 2(1), 7–28 (2000)

Hall, E.T.: Beyond Culture, New York (1989)

Heckmann, D.: Ubiquitous User Modeling. PhD dissertation, Universität Saarbrücken, Saarbrücken, Germany, [online] (2005), available at http://w5.cs.uni-sb.de/publication/file/ 178/ Heckmann05Diss.pdf

Heier, H., Borgman, H.: Knowledge Management Systems Spanning Cutures: the Case of Deutsche Bank's HRbase. In: Proceedings of the 10th ECIS, Gdansk, Poland (June 6-8, 2002)

Herring, S., Estrada, Z.: Representations of Indigenous Language Groups of North and South America on the World Wide Web: In Whose Voice? In: Sudweeks & Ess, pp. 377–380 (2004)

Herring, S., Paolillo, J., Clark, B., Kouper, I., Ramos-Vielba, I., Scheidt, L.A., Stoerger, S., Wright, E.: Linguistic Diversity and Language Networks on LiveJournal. In: Proceedings of the INSNA Sunbelt Conference, Vancouver, Canada (2006)

Hofstede, G.: Cultures and Organizations: Software of the Mind. McGraw-Hill, London (1991)

Hongladarom, S.: Global Culture, Local Cultures, and the Internet: The Thai Example. In: Sudweeks & Ess, pp. 187–201 (2004)

Internet World Stats, Internet World Users by Language. (2007), http://www.internetworldstats. com/stats7.htm

Jameson, A.: Modeling both the context and the user. Personal Technologies 5(1), 29–33 (2001)

Jarvenpaa, S.L., Tractinsky, N.: Consumer Trust in an Internet Store: A Cross-Cultural Validation. Journal of Computer-Mediated Communication, 5(2), (1999), [online] available at http://jcmc.indiana.edu/vol5/issue2/jarvenpaa.html

Kobsa, A., Teltzrow, M.: Impacts of Contextualized Communication of Privacy Practices and Personalization Benefits on Purchase Behavior and Perceived Quality of Recommendation. In: Van Setten, M., McNean, S., Konstan, J. (eds.) Beyond Personalization 2005: A Workshop on the Next Stage of Recommender Systems Research (IUI 2005), San Diego, CA, USA, pp. 48–53 (2005)

Kralisch, A.: The Impact of Culture and Language on the Use of the Internet: Empirical Analyses of Behaviour and Attitudes. PhD dissertation, Humboldt University, Berlin. (2006), Published at http://edoc.hu-berlin.de/docviews/abstract.php?id=27410

Kralisch, A., Berendt, B.: Cultural Determinants of Search Behaviour on Websites. In: Evers, V., del Galdo, E., Cyr, D., Bonanni, C. (eds.) Proc. of the 6th IWIPS, Product & Systems Internationalisation, Vancouver, Canada, July 8-10, pp. 61–74 (2004)

Kralisch, A., Berendt, B.: Language-sensitive Search Behaviour and the Role of Domain Knowledge. New Review of Multimedia and Hypermedia: Special Issue on Minority Language, Multimedia and the Web 11(2), 221–246 (2005)

Kralisch, A., Berendt, B., Eisend, M.: Impact of Culture on Website Navigation Behaviour. In: Proc. of 11th Int. Conf. on Human-Computer Interaction, Las Vegas, NE (July 22-27, 2005)

Kralisch, A., Köppen, V.: The Impact of Language on Website Use and User Satisfaction: Project Description. In: Bartmann, D., et al. (eds.) Proc. of the 13th ECIS, Regensburg, Germany (May 26-28, 2005)

Kralisch, A., Mandl, T.: Intercultural Aspects of Design and Interaction with Retrieval Systems. In: Proc. of 11th Int. Conf. on Human-Computer Interaction, Las Vegas, NE, (July 22-27, 2005)

Kralisch, A., Mandl, T.: Barriers to Information Access across Languages on the Internet: Network and Language Effects. In: Proceedings of the 39th Hawaii International Conference on System Sciences (HICSS-39), Poipu, HI, USA, January 4-7, IEEE Computer Society, Los Alamitos (2006)

Mafu, S.: From the Oral Tradition to the Information Era: The Case of Tanzania. International Journal of Multicultural Societies 6(1), 53–78 (2004)

Marcus, A., West Gould, E.: Cultural Dimensions and Global Web User-Interface Design: What? So what? Now what? In: Proc. of the 6th Conf. on Human Factors and the Web (HFWeb), Austin, TX, USA (June 19, 2000)

Markova, M.: Barriers for Informationo Technology Adoption in Post-Soviet Central Asia. In: Sudweeks & Ess, pp. 277–281 (2004)

McEneaney, J.E.: Graphic and numerical methods to access navigation in hypertext. Int. J. Hum.-Comput. Stud. 55(5), 761–786 (2001) http://www.oakland.edu/~mceneane/ijhcs.pdf

Milberg, S.J., Smith, H.J., Burke, S.J.: Information Privacy: Corporate Management and National Regulation. Organizational Science 11, 35–37 (2000)

Morden, T.: Models of National Culture – A Management Review. Cross Cultural Management 6(1), 19–44 (1999)

Palfreyman, D., Al-Khalil, M.: A Funky Language for Teenzz to Use: Representing Gulf Arabic in Instant Messaging. Journal of Computer-Mediated Communication, 9(1), (2003), http://jcmc.indiana.edu/vol9/issue1/palfreyman.html

Panyametheekul, S., Herring, S.: Gender and Turn Allocation in a Thai Chat Room. Journal of Computer-Mediated Communication, 9(1), (2003), jcmc.indiana.edu/vol9/issue1/panya_herring.html

Pargman, D., Palme, J.: Linguistic Standardization on the Internet. In: Sudweeks & Ess, pp. 385–388 (2004)

Van Dyke Parunak, H.: Hypermedia Topologies and User Navigation. In: Proceedings of ACM Hypertext '89, pp. 43–50 (1989)

Phillipson, R.: Linguistic Imperialism. Oxford University Press, Oxford (1992)

Phillipson, R., Skutnabb-Kangas, T.: Linguistic Imperialism. In: Mesthrie, R. (ed.) Concise Encyclopedia of Sociolinguistics, pp. 570–574. Elsevier Science, Oxford (2001)

Reeder, K., Macfadyen, L.P., Chase, M., Roche, J.: Falling through the (Cultural) Gaps? Intercultural Communication Challenges in Cyberspace. In: Sudweeks & Ess, pp. 123–134 (2004)

Siala, H., O'Keefe, R.M., Hone, K.: The Impact of Religious Affiliation on Trust in the Context of Electronic Commerce. Interacting with Computers 16(1), 7–27 (2004)

Sudweeks, F., Ess, C. (eds.) Proceedings of the Fourth International Conference on Cultural Attitudes towards Technology and Communication (CATaC), School of Information Technology Murdoch University, June 27 - July 1, Karlstad, Sweden (2004)

Tseliga, T.: It's all Greeklish to me!:" Linguistic and sociocultural perspectives on roman-alphabeted Greek in asynchronous computer-mediated communication. In: Danet, B., Herring, S. (eds.) The Multilingual Internet: Language, Culture and Communication Online, Ch. 5, Oxford University Press, New York (2007)

Warschauer, M.: Technology and Social Inclusion: Rethinking the Digital Divide. MIT Press, Cambridge (2003)

Wei, C., Kolko, B.: Resistance to Globalization: Language and Internet Diffusion Patterns in Uzbekistan. New Review of Multimedia and Hypermedia 11(2), 205–220 (2005)

Wheeler, D.: Old Culture, New Technology: A Look at Women, Gender and the Internet in Kuwait. In: Ess, C., Sudweeks, F. (eds.) Culture, Technology, Communication: Towards an Intercultural Global Village, Suny, Albany (2001)

Wodak, R., Wright, S.: The European Union in Cyberspace: Democrative Participation via Online Multilingual Discussion Boards? In: Danet, B., Herring, S. (eds.) The Multilingual Internet: Language, Culture and Communication Online, Oxford University Press, New York (2004)

Yeo, A., Loo, W.: Identification and Evaluation of Classification Schemes: A User-Centred Approach. In: Evers, V., del Galdo, E., Cyr, D., Bonanni, C. (eds.) Proceedings of the 6th IWIPS (2004)

Aspect-Based Tagging for Collaborative Media Organization

Oliver Flasch, Andreas Kaspari, Katharina Morik, and Michael Wurst

University of Dortmund, AI Unit
{flasch,kaspari,morik,wurst}@ls8.cs.uni-dortmund

Abstract. Organizing multimedia data is very challenging. One of the most important approaches to support users in searching and navigating media collections is collaborative filtering. Recently, systems as flickr or last.fm have become popular. They allow users to not only rate but also tag items with arbitrary labels. Such systems replace the concept of a global common ontology, as envisioned by the Semantic Web, with a paradigm of heterogeneous, local "folksonomies". The problem of such tagging systems is, however, that resulting taggings carry only little semantics. In this paper, we present an extension to the tagging approach. We allow tags to be grouped into aspects. We show that introducing aspects does not only help the user to manage large numbers of tags, but also facilitates data mining in various ways. We exemplify our approach on Nemoz , a distributed media organizer based on tagging and distributed data mining.

Keywords: Data mining, social bookmarking, multimedia data.

1 Introduction

Networks allow users to access information in a transparent, location-independent way. The ability to exchange any kind of data and information, independent of ones current geographical location, paves the way for patterns of cooperation that have not been possible before. The collaborative Wikipedia contains more articles than the Encyclopedia Britannica. On the Usenet, each day 3 Terabyte of information is exchanged in more than 60,000 discussion groups. However, in order to make full use of these possibilities, search engines are not enough, but *intelligent mediation* is needed. Mediation refers to the task of making information provided by one user accessible and beneficial for other users. One option is to enrich resources of the Internet with more semantics. The most ambitious project towards this goal is the Semantic Web. While this approach would allow for a better semantic description, it faces considerable acceptance problems. Assuming a global semantic model (a global view of the world) does not reflect the very subjective way in which users handle information. Annotating resources using Semantic Web formalisms requires too much time and is rewarded only indirectly and after quite a while. Hence, an alternative option has recently been proposed, the Web 2.0, where tagging is performed individually at the distributed local sites without reference

B. Berendt et al. (Eds.): WebMine 2006, LNAI 4737, pp. 122–141, 2007.

to a global ontology. Examples are tagging systems such as flickr[1], del.icio.us[2], or last.fm[3]. They allow users to assign arbitrary tags to content. Formal ontologies are replaced by so-called "folksonomies", that do not depend on common global concepts and terminology. The ability to freely choose concepts used for annotation is the key that led to the high acceptance of these systems by users. One drawback of these approaches is, however, that automatic analysis of the resulting structures is difficult because they are usually extremely ambiguous. Web 2.0 tagging trades in semantic information for the ease of use.

Intelligent mediation can be investigated looking at recommender systems [1,2,3]. They are a cornerstone of the current Internet and can be found in a broad range of applications ranging from online stores as Amazon, over music organizers to web site recommenders. Recommendations may be based on knowledge about the domain or on similarity of users. The knowledge-based approach can be illustrated by the Pandora system of Tim Westergren's Music Genome Project. Pandora exploits careful expert annotations of music which are based on a music-theoretic ontology and matches them with user feedback for the recommendation of music, presented as on-line Internet radio. The collaborative approach can be illustrated by Amazon, where the shopping behavior of customers is observed and the overlapping items in the baskets of several users is used as a similarity measure of users. The associated items which are not yet overlapping are recommended.

Currently large amounts of multimedia content are stored at personal computers, MP3 devices, and other devices. The organization of these collections is cumbersome. First, the users have a different view of their collection depending on the occasions in which they want to use it. For instance, for a party you retrieve rather different music than for a candle-light dinner. Also places are related to views of the collection. In the car you might prefer rather different music than than at the working place. Moreover, a good host might play music for a special guest which he himself doesn't like usually. Hence, there is not one general preference structure per person, but several preference structures of the same person. Of course, some people share their views. However, it might well happen that one person's favorite songs for car-driving are best liked by another person while cleaning the house, where for car-driving this person wants different music. We generalize these observations to *aspects* of structuring collections. Second, flat structures like the one used by collaborative filtering are not sufficient. There, the correlation of items is taken into account, but no further structures. This leads to the presentation of a collection in terms of a table (like in iTunes) or just lists. Instead, we would like to organize a collection in terms of hierarchies so that browsing through the collection becomes easy, and the user gets a nice overview of the collection.

In this paper we present ongoing work on the project Nemoz , a collaborative music organizer based on distributed data and multimedia mining techniques.

---

[1] See http://flickr.com
[2] See http://del.icio.us
[3] See http://last.fm

Distributed organization of music is an ideal test case for ubiquitous data mining. First, the management of multi-media data is a very hard task because corresponding semantic descriptions depend on highly social and personal factors. Second, media collections are usually inherently distributed. Third, multi-media data is stored and managed on a large variety of different devices with very different capabilities concerning network connection and computational power. In this paper, we show that some of the problems connected with the automatic analysis of user-created tags can be solved by allowing to group tags into aspects. We investigate, how to tailor the representation such that it supports the personalized services provided to users as well as distributed data mining techniques.

The rest of this work is structured as follows: Section 2 introduces the Nemoz Project. Section 3 introduces the idea of multi-aspect tagging and discusses several possible restrictions, their utility and their implications. In section 4 we analyze how aspect-enriched tagging structures can be exploited by distributed data mining methods. In section 5 we present a prototypical implementation of a network multimedia organizer based on these ideas. In section 6 we close with a conclusion.

2   The Nemoz Project

One of the most successful approaches to information and multimedia organization is tagging. Users tag items, such as webpages or music, with arbitrary chosen textual descriptions. These descriptions might describe global properties, as the name of the item, or highly personal, such as "music for work". Tags can be shared by globally attaching them to items. Other users can then exploit these tags to search for specific items or to browse them. Also, items are shown in conjunction with all tags that were assigned to them. In this way, a user can survey under which tags an item was classified by other users. Social bookmarking systems are widely used for two reasons. The first one is rather selfish. Users want to organize their own item collection in a convenient and intuitive way on the Internet (across different hardware or software platforms). The second one is benevolent. Users want to contribute to the common knowledge about web resources as to help other users to find interesting information [Hammond/etal/2005a].

While tagging is used in many information and multimedia organization systems, there are still many open challenges to make these systems more user friendly.

1. Personal tag collections quickly become hard to manage as the number of tags grow. This is especially the case, if users apply tags that represent different aspects of the underlying items, e.g. "rock" denoting a genre, "party" denoting a genre or an occasion and "driving" denoting an occasion.
2. While tagging some items is fun for most users, tagging a complete item collection (e.g. all audio files one owns) is often not perceived as pleasure.

Especially as the collection of items constantly grows, users must keep their assignment of tags to items always up to date.

3. Tag collections evolve over time. Methods that support users in tagging items must respect this incremental nature of personal tag collections.
4. Finally, most current social bookmarking systems are based on a client-server architecture, allowing to analyze all data at a central node. On the other hand, many current and future scenarios will be based on distributed ad hoc cooperations in which no central node is available.

The Nemoz Project develops new interaction and data mining methods that cope with these challenges. It envisions a set of innovative features and functions that future distributed media organization applications should provide, and develops methods to achieve them. Beside the usual functionality of social bookmarking systems, we consider the following additional functions:

1. Users might not only tag items, but also group tags into a sets of aspects. Typical aspects would be "genre", "country", "style" or "mood". This allows to structure items according to a desired aspect and allows to filter tags by an aspect making them much easier to manage, especially as the number of tags grows.
2. Users may only tag a small number of items and then leave it to the system to assign tags to the rest of the items. This is especially useful as items arrive one-by-one and are automatically added to the existing tag structures. The fact that tags are arranged into aspects is essential at this point, as we do not have to deal with one classification problem per tag, but one per aspect, which is usually much more efficient and intuitive.
3. Often users first collect items and then end up with a large number of unclustered or only poorly clustered items. A media organization system should be able to recommend several alternative, sound tag structures for these items automatically. The user can then choose from these structures and possibly modify them. Returning several alternatives is crucial, as structuring is an explorative task and the system usually cannot guess, which structure will fit the user's needs best. Also, even if a set of items is already tagged, the user might want to explore additional possibilities to structure these items.
4. As users explore item collections, e.g. items owned by other users, it is very convenient to visualize and navigate these items using one's own tags, instead of the tags assigned by someone else. This can be achieved by assigning one's own tags temporarily to items in any collection by classification.

All of this functionality should be provided in a fully distributed way, as media collections are not only stored on personal computers with a highspeed network connection, but also on mobile devices, such as cell phones, portable players, etc.

In the following two sections we show how this functionality can be achieved by a combination of a new interaction method called "aspect-based tagging" and by corresponding data mining methods. These methods are based on the idea of distributed, collaborative data mining. To support a current user in structuring her media collection, we exploit tags and information provided by other users in a fully distributed way. This leads to the notion of "collaborative classification" and "collaborative structuring" and will be described in detail in section 4.

3   Multi-aspect Tagging

One of the major challenges in enabling distributed, collaborative media organization is to find an appropriate representation mechanism. While developing the Nemoz system, we found a set of requirements that a representation mechanism must fulfill to be well-suited for distributed, collaborative media organization. We will show that neither current Semantic Web approaches nor popular Web 2.0 tagging approaches fulfill these requirements. This is the point of departure for our aspect-based tagging approach.

1. *No explicit coordination*
   The growth of the Internet can be attributed largely to its loosely coupled character. If, for instance, every owner of a site would have to agree, that someone links to her site, the Internet would probably not have grown as fast as it did, nor would approaches as link analysis be as powerful as they are. We therefore require that a representation mechanism must not depend on explicit coordination among users.

2. *Co-existence of different views*
   Often, users do not agree on how to structure a certain set of items. It is therefore essential, that different representations of the same items may co-exist. In the extreme, each user should be allowed to create views completely independently of all other users. This allows for bottom-up innovation, as each user is capable of creating novel views. Which views become popular should emerge automatically, just like popular web-pages emerge automatically, as many other pages link to them.

3. *Support for data mining and mediation*
   While using loosely coupled representations is very attractive, the question remains how to derive useful information from such heterogeneous views and to allow users to profit from what other users did. A representation mechanism should therefore allow for the successful application of data mining and mediation methods.

4. *Efficiency*
   Relevant operations must be executable efficiently. For media management, the most important operations are the retrieval of items, basic consistency checks and the application of data mining methods, such as automatic classification.

5. *Manageability*
   The representation mechanism should be such, that it is easy for the user to overview and maintain the knowledge structures she created.

6. *Ubiquitous environments*
   The mechanism must be applicable in highly distributed environments. It must not expect, that all nodes are connected to the network all the time. Also, distributed data mining and retrieval methods must be applicable, such

that the overall effort in communication time and cost is low, as media data is often organized on computationally poor devices connected by a loosely coupled network (such as p2p or ad hoc networks).

On the other hand, we think that other properties of knowledge representation mechanisms, especially as developed by the AI community, are not overly relevant for media organization. First, the representation of complex relationships is not of essential importance. Regular users are often not capable of dealing with such complex relationships (the large majority of Google users never even applied simple logical operators in their search requests). Also, complex relations are only seldom contained in the domain in question. Most properties can be simply expressed by pairs of attribute and value (artist, year of publication, ...). Furthermore, logical inference is often not useful, as most users express their knowledge rather ad hoc and do not even accept logical entailment of what they expressed. We do not claim however, that these properties are irrelevant in general, we only claim that they are not relevant for media organization.

The most important representation mechanism for Internet resources is the Semantic Web. It is based on first order logic based representation mechanism. Given the above requirements, the Semantic Web is not well suited as representation mechanism for media organization. It is quite complex and requires explicit coordination among users. The co-existence of views and emerging views are not directly supported. Also, as the representation mechanism is quite powerful, operations may become inefficient. It is based on logical entailment and is often not comprehensible for regular users. Finally, as it is usually based on explicit coordination, it can be hard to implement in a ubiquitous environment.

Recently, new applications emerged under the Web 2.0 paradigm. Systems as flickr or del.icio.us allow users to annotate items with arbitrary chosen tags. Such tags complement global properties, e.g. artist, album, genre, etc. for music collections used by traditional media organizers. In contrast to these global properties, many user-assigned tags are *local*, i.e. they represent the personal views of a certain user not aiming at a global structure or semantic. These systems allow for multiple and emerging views, do not require any coordination and are very easy to implement in an ubiquitous environment. A major drawback is, that tag structures tend to be chaotic and hard to manage. In order to offer its users a more structured overview of their tag collection, del.icio.us recently introduced a new feature called *tag bundles* which allows arranging tags into named groups. The major drawback is that tag bundles unlike tags are not shared among the users of the system and cannot be browsed. They are not part of the knowledge representation formalism and serve only as a means of organizing a user's personal collection of tags.

Another problem of tag structures is that they are not really well suited for data mining, which is a prerequisite for collaborative media organization.

In the following we show, that we can weaken these problems by introducing a knowledge representation formalism designed to support the concept of aspects.

Folksonomies emerging from popular social content services like last.fm or flickr constitute a large source of information. By virtue of compatibility, our

formalism makes this information available for ontology-based knowledge discovery. Representing information from existing services consistently in one formalism enables us to create "mash-ups" of these services, i.e. to join data from
multiple sources. This possibility is a defining trait of Web 2.0 applications. By
integrating into the existing Web 2.0, new applications avoid the dilemma of a
"cold start". BibSonomy[4] (see also [4]), a collaborative bookmark and publication sharing system, includes DBLP data in this fashion.

In the following subsections, we describe a variation of the folksonomy representation formalism given in [5], supporting the concept of aspects.

3.1   Basic Entities and Concepts

The basic entities in our formalism are *users*, *items* (songs), *categories*, and
aspects:

Definition 1. *(Domain sets)*

$$U = \{u_1, \ldots, u_l\} \ \text{(User Identifiers)}$$
$$I = \{i_1, \ldots, i_n\} \quad \text{(Item Identifiers)}$$
$$C = \{c_1, \ldots, c_m\} \ \text{(Category Identifiers)}$$
$$A = \{a_1, \ldots, a_k\} \ \text{(Aspect Identifiers)}$$

Instead of storing these entities directly, we distinguish between abstract, opaque
entity identifiers and entity representations. This distinction is motivated by the
"Representational State Transfer" [6] paradigm of the World Wide Web, to which
our formalism adheres to. In the rest of this work, we will only deal with abstract
entity identifiers in the form of URNs.

In the following paragraphs, we describe the concepts of our formalism as a
series of extensions to the Web 2.0 tagging model. In this model, users annotate
a common set of items with tags. We represent tags by category identifiers. Links
between items and tags are called $\mathcal{IC}$-Links:

Definition 2. *($\mathcal{IC}$-Link Relation)*

$$\triangleright_{IC}: \subseteq I \times C.$$

Our concept of a category extends the Web 2.0 tagging model by explicitly
allowing "categories of categories", thereby enabling the representation of hierarchical structures akin to first order logic and description logics [7]:

Definition 3. *($\mathcal{CC}$-Link partial order)*
The $\mathcal{CC}$-Link partial order is a relation

$$\preceq_{CC}: \subseteq C \times C$$

which satisfies the following axiom:

$$c \preceq_{CC} c' \Rightarrow ext(c) \subseteq ext(c') \ \text{where} \ c, c' \in C, \tag{1}$$

where $ext(c)$ is the item extension of a category c.

---

[4] Online at http://www.bibsonomy.org

Note that in our formalism, the fact that $\text{ext}(c) \subseteq \text{ext}(c')$ does *not* imply that $c \preceq_{CC} c'$. We will motivate this design decision by an example: Consider a user whose music library contains very little jazz, all by Miles Davis. Our formalism would not force this user to accept the rather nonsensical identification of jazz and Miles Davis implied by the identity of the extension sets. If this identification actually reflects the user's opinion, she is still free to declare it explicitly.

Our formalism allows the user to organize categories further by grouping them into aspects:

Definition 4. *(CA-Link Relation)*

$$\blacktriangleright_{CA} : \subseteq C \times A$$

Typical examples for aspects from the music domain are "genre", "mood", "artist" and "tempo". The addition of aspects enables, among other things, the extraction of corresponding taxonomies, as described in section 4.

The usefulness of aspects has several facets. First, hierarchical category structures tend to become unmanageable when growing in size. Aspects enable the user to create complex structures to organize her items and simultaneously maintain clarity. Consider a user, who uses del.icio.us to organize her hyperlinks. With a great number of tags, retrieving one such link becomes more and more complicated. Grouping tags/categories into aspects eases this task considerably. Second, aspects can be used for filtering large category structures. Filtering means restricting the visible fraction of these structures to a specific topic. A limited variant of this notion is implemented in the iTunes media organizer, where the user can select a genre or an artist she wants to browse. Our framework enables the user to browse her items by arbitrary aspects. Third, aspects implicitly define a similarity measure on items that can be used to realize aspect-based structuring and visualization.

All links are considered as first class objects, facilitating the implementation of the formalism in a distributed environment.

3.2 Users and Ownership

In our formalism, entities are "ownerless", only links are owned by users:

Definition 5. *(Link-ownership Relation)*

$$\triangleright_O : \subseteq (\triangleright_{IC} \cup \preceq_{CC} \cup \blacktriangleright_{CA}) \times (\mathcal{P}(U) \backslash \emptyset)$$

Each link must have at least one owner. It may have multiple owners, if it has been added independently by multiple users. This is why we wrote the power set of users, $\mathcal{P}$. The Link-ownership Relation can easily be embedded in the IC-, CC- and CA-Link Relations, we chose not to do so to facilitate implementation in an object-oriented language. This decision should be reconsidered if this formalism where to be implemented using a relational database system.

Item ownership is not a first class concept in our formalism. Nonetheless, our prototypic implementation (Nemoz) provides a notion of item ownership: An item (i.e., a song) is said to be "owned" by a User, if this User possesses a representation of this item (i.e., an audio file of this song) stored on her local machine.

Our user concept comprises human users as well as intelligent agents. An agent acts on behalf of a human user, but has an identity of its own. For example, the "intelligent" operations (i.e. clustering and classification) of Nemoz (see section 2) have been modeled using such agents. Each time an intelligent operation is triggered by a user, an agent user is created that performs the operation and adds the resulting links to the knowledge base. Our design gives the user control over the effects of these operations by clearly distinguishing between automatically generated and manually entered knowledge. An automatically generated link may be promoted to a user-approved link by changing the link ownership from an agent to its client user. The effects of an intelligent operation may be canceled by deleting the responsible agent. By keeping automatically generated knowledge in an ephemeral state until it has been approved by the user, we hope to tame the sometimes frustrating effects of a poor performing intelligent operation.

3.3   Nemoz Knowledge Bases

With the preliminaries in place, we are now able to define our notion of an aspect-enriched tagging structure:

Definition 6. *(Nemoz Knowledge Base) A* Nemoz Knowledge Base KB_{Nemoz} *is defined as an 8-tuple:*

$$KB_{Nemoz} := (I, C, A, U, \rhd_{IC}, \preceq_{CC}, \blacktriangleright_{CA}, \rhd_O),$$

which satisfies the following axioms:

$$\forall c \in C.\exists i.(i, c) \in \rhd_{IC} \tag{2}$$
$$\forall a \in A.\exists c.(c, a) \in \blacktriangleright_{CA} . \tag{3}$$

These axioms ensure that all categories and aspects in a Nemoz Knowledge Base are not empty, a property we will refer to as *supportedness*. Supportedness implies that all categories and aspects have "extensional support", which is favorable from a machine learning perspective as well as from a user perspective.

Constraining the definition of a Nemoz Knowledge Base, we can describe tagging systems as well as some description logics-based formalisms.

An obvious restriction leads to *flat Nemoz Knowledge Bases*, that disallow hierarchically structured categories:

Definition 7. *(flat* Nemoz Knowledge Base) A flat Nemoz Knowledge Base $KB_{Nemoz/flat}$ *is defined as a* Nemoz Knowledge Base *without CC-Links* ($\preceq_{CC} = \emptyset$), *described as a 7-tuple:*

$$KB_{Nemoz/flat} := (I, C, A, U, \rhd_{IC}, \blacktriangleright_{CA}, \rhd_O).$$

A flat Nemoz Knowledge Bases is an aspect-enriched tagging system. These systems offer the benefits of aspects without the complexity of hierarchical category structures.

A further restriction leads to simple tagging systems:

Definition 8. *(Tag Knowledge Base) A Tag Knowledge Base KB_{tag} is defined as a flat Nemoz Knowledge Base without aspect identifiers $(A = \emptyset)$ which implies an empty $\mathcal{CA}$-Link Relation $(\blacktriangleright_{CA} = \emptyset)$. Thus, a Tag Knowledge Base can be described as a 5-tuple:*

$$KB_{tag} := (I, C, U, \rhd_{IC}, \rhd_O).$$

A Tag Knowledge Base is a special case of a Nemoz Knowledge Base and may be seamlessly enriched by hierarchical categories or aspects. At the same time, each Nemoz Knowledge Base may be stripped down to a Tag Knowledge Base in a trivial manner. This flexibility enables simple inter-operation with existing knowledge bases of the Web 2.0.

A direct advantage of aspects is that the user is not confronted with a large number of tags, but with only some aspects that can be used to select subsets of tags. This essentially eases the visualization and maintenance of tag structures.

In the next section, we show that users profit from aspects yet in another way. The resulting structures are much better suited for data mining, which is the basis for the collaborative functionality of the system.

4   Aspect-Based Multimedia Mining

The first function envisioned in Sec. 2, the grouping, filtering and visualizing of tags into aspects was described in the last section. The remaining functions are based on data and multimedia mining. More specifically, tagging new items with existing tags or visualizing item collections using one's own tags can be seen as classification tasks. Finding (alternative) tag structures for so far untagged items can be seen as a clustering task.

While there are many algorithms for classification and clustering, applying them to multimedia data is still very challenging. In this section we propose several approaches that exploit the fact, that many users face similar data mining tasks and that we can share information among these tasks to optimize the data mining process.

4.1   Multimedia Mining

Applying data mining methods to the field of personal media management offers many new opportunities. Typical applications include the classification of music items according to predefined schemes like genres [8,9], automatic clustering and visualization of audio clips [10,11], recommendations of songs [12], as well as the automatic creation of playlists based on audio similarity and user feedback [13,14].

A key issue in all these approaches is the representation of the underlying items. Confronted with music data, machine learning encounters a new challenge of scalability. Music databases store millions of records and each item contains up to several million values. In addition, the shape of the curve defined by these values does not express the crucial aspect of similarity measures for musical objects. The solution to overcome these issues is to extract features from the audio signal which leads to a strong compression of the data set at hand. Many manually designed audio features extracted from polyphonic music have been proposed for extracting features from audio data [15,16].

However, it turns out that optimal audio features strongly depend on the task at hand [17] and the current subset of items [11]. It is not very likely that a feature set delivering excellent performance on the separation of classical and popular music works well also for the separation of music structured according to occasions. This problem already arises for high-level structures like musical genres and is even aggregated due to the locality induced by personal structures. One possibility to cope with this problem is to learn an adapted set of features for each learning task separately [18,19]. These approaches achieve a high accuracy, but are computationally very demanding and not well suited for real time processing.

If there would exist one complete set of features, from which each learning task selects its proper part, this problem could be reduced to feature selection. However, there is no tractable feature set to select from. The number of possible feature extractions is so large – virtually infinite – that it would be intractable to enumerate it.

Beside these problems, emotional or socio-cultural aspects of music can hardly be expressed by feature values at all. Clustering schemes merely using audio features as basis of a similarity measure will fail for this reason. Such aspects have a significant influence on how people structure and perceive their music [20].

This is especially a problem for clustering multimedia data, as different features sets lead to completely different clustering solutions.

In the following we first describe a collaborative approach to feature construction for classification that is efficient and still achieves high accuracy. Then we discuss a collaborative approach to clustering.

4.2   Collaborative Classification

One of the most important assistant functions for the user is to tag items automatically based on examples. In our case, these labeled examples are items already annotated with tags. The supportedness condition ensures that there actually is an example for each tag. As discussed above, an advantage of aspect-based tagging is, that we face a classification problem per user and aspect. In traditional tagging system we would face a binary classification problem per user and tag.

In principle, this task can be achieved by any state-of-the art (hierarchical) classification algorithm. However, to be successful, we have to select a set of features $X_t \subseteq \mathbf{X}$ that are well suited for this task from a possibly infinite set of possible features $\mathbf{X}$. While algorithms as proposed in [18] allow for this kind of feature construction, these methods are by far too inefficient to be applied in a system that requires "real time" responses.

If there were only a single isolated classification task, there would not be much to do about this problem. In a networked media setting, we have not a single task, but a constantly growing set of tasks T. We can assume that for some of these tasks, optimal features were already identified. Instead of searching for adequate feature $X_{t'}$ for a new task t', we can try to transfer successful features from the set of tasks for which such features are already known. This allows us to achieve both: a high accuracy due to an adequate feature set and low response times, as we do not need to construct the features anew, but simply "reuse" them.

Classification tasks defined by user assigned tags can differ in any possible way, concerning the features they demand. Therefore, the transfer of features among tasks must be selective. Features should only be shared among similar tasks. While there is no guarantee that any two tasks are similar, we assume that this is the case, as many users organize their music in similar ways.

One of the first applications of selective information transfer among learning tasks is presented in [21]. This approach requires however, to cross-apply feature sets to the original example sets to assess the similarity of the tasks. In a networked scenario, this is much to inefficient, as it requires to share complete example sets. In [22] we proposed a novel approach that determines the similarity of tasks based on a scalar vector only. This method can easily be applied in distributed settings, in which we do not have a good internet connection. It is therefore very well suited for ubiquitous applications.

As this method is effectively based on sharing information among different user defined classification tasks, we denote it as "collaborative classification". It is local, as it does not try to find global annotations or models. The idea is rather to regard a large number of learning tasks independently of each other, still sharing a maximum of information among these tasks. This paradigm differs from traditional web mining paradigms, that try to find global annotations of resources. We think that this local paradigm to web mining will become more and more important as the number of such user created data (bookmarks, personal taxonomies, etc.) on the Web grows.

4.3 Collaborative Structuring

Comparing classification tasks allows us to reuse features among them to make the system efficient and accurate at the same time. If the task is to structure items that are not yet structured, we face a clustering task instead. Can we achieve a transfer of features or information among clustering tasks as well?

Depending on the underlying feature set, the same set of items can be clustered very differently. Some approaches allow users to state constraints on how items should be clustered [23]. In many media organization applications, users are not patient enough to go through such a often iterative process.

We therefore propose an approach that represents a clustering problem only by the items to be clustered. Instead of clustering the items directly, we first search for existing aspects (and thus a set of tags) that already covers the items to be clustered or at least some of them. An aspect covers an item, if a tag belonging to this aspect is assigned to the item. As tags belonging to an aspect are created semi-automatically, they can be regarded as "sound" clustering of all items the aspect covers.

We can therefore apply these tags to the items to be clustered. All items not yet covered directly can be added by classification. An appropriate feature set for classification can be found using the method described in the last section. This enables us to use localized features for clustering as well.

Instead of returning only a single clustering, the system can also search for several aspects that cover the items in question. This then results in several alternative solutions, from which the user can choose.

If the set of items is heterogeneous, it might only be possible to cover it by combining several aspects and thus tag structures. This leads to the idea of a bag of clustering, as used in the LACE algorithm [24].

We denote this approach as "collaborative structuring", as it allows users to cluster items in a collaborative way. As clustering tasks are represented extensionally by the set of items that are or should be clustered, queries used for the LACE algorithm can all be reduced to simple search for items. This kind of search is very well-supported by current p2p technology on which these methods can be based.

The basic paradigm of collaborative clustering is the same as the one of collaborative classification. We assume a large number of local clustering tasks that are in general not related to each other. Still we can assume that many personal tag structures resemble each other to some extend. This allows us to share information among these clustering tasks.

From a point of view of the user, this is very attractive, as it allows to position herself between two extremes. Either she can simply use the tag structure recommended to her by the LACE algorithm, or she can create her own tag structure completely from scratch, enriching the system with a new point of view.

5   The **Nemoz** Prototype

Together with a group of students we have developed Nemoz [5] as a framework for studying collaborative music organization. Nemoz is made for experimenting with intelligent functionality of distributed media organization systems. Of course, the basic functions of media systems are implemented: download and import of songs, playing music, retrieving music from a collection based on given meta data, and creating play lists.

---

[5] Nemoz is available as an open source project at http://nemoz.sf.net

5.1   Data Model and Architecture

First, the Nemoz data model contains an implementation of the knowledge representation formalism described in section 3. Standard music metadata (performer, composer, album, year, duration of the song, genre, and comment) is automatically extracted from ID3-Tags[6] and represented in this formalism: Each metadata attribute type is mapped to an aspect, each metadata value is mapped to a category of this aspect. For example, after being added to the system, the item "So What" by the artist "Miles Davis" will be contained in the category "Miles Davis", which will be linked to the aspect "Artist". Second, the Nemoz data model also contains a mapping from items to features which are extracted from raw sample data.

Communication among (W)LAN nodes via TCP and UDP is supported by a network service.

A collection can be organized using several aspects in parallel. Based on this data model Nemoz offers several (intelligent) functions:

- Category tags can be assigned to arbitrary sets of items. If a category tag does not exist at assignment time, it will be created.
- A category tag itself can be tagged, forming a hierarchy of categories.
- Category tags can be grouped into arbitrary aspects.
- Category tags can be automatically assigned to new items.
- Users can search for similar aspects and categories in the network.
- Users can search for music similar to a selected song or group of songs.
- Tag structures can be automatically enhanced through the tags of other users.

By means of these functions, each user may create arbitrary, personal classification schemes to organize her music. For instance, some users structure their collection according to mood and situation, others according to genre, etc. Some of these structures may overlap, e.g., the blues genre may cover songs which are also covered by a personal category "melancholic" of a structure describing moods.

Intelligent functions are based on the principles described in section 4. Nemoz supports the users in structuring their media objects while not forcing them to use a global set of concepts or annotations. If an ad hoc network has been established, peers may support each other in structuring.

By recommending tags and structures to other users, we establish emerging views on the underlying space of objects. This approach naturally leads to a social filtering of such views. If someone creates a (partial) tag structure found useful by many other users, it is often copied. If several tag structures equally fit a query, a well-distributed tag structure is recommended with higher probability. This pushes high quality tag structures and allows to filter random or non-sense ones. While the collaborative approach offers many opportunities, audio features can still be very helpful in several ways. The most important is that they allow

---

[6] See http://www.id3.org

to replace exact matches by similarity matches. This is essential when dealing with sparse data, i.e. when the number of objects in the tag structure is rather small.

5.2   User Interface

Describing an intuitive user interface for Nemoz poses a challenge that exceeds the scope of this text. It may be the case that no single user interface concept is optimal for all use cases of a collaborative media organizer like Nemoz . We implemented several different user interfaces to explore the large space of possibilities. All these user interfaces are implemented as plugins to a common Nemoz Kernel, and can operate concurrently on a common data model. This architecture simplifies experimentation with several contrasting user interface concepts considerably.

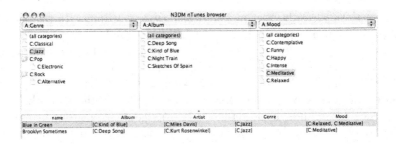

Fig. 1. Filtering browser (extract)

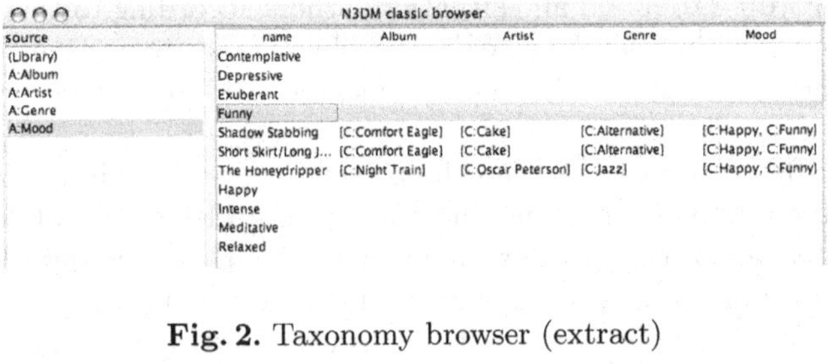

Fig. 2. Taxonomy browser (extract)

Figure 1 shows the filtering browser, which is inspired by well known music organizers such as Apple's iTunes [7] or Nullsoft's Winamp [8]. This interface gives the user a filtered view of all the items in her media library, which can be successively refined by applying up to three aspects filters. In contrast to existing

---

[7] http://www.apple.com/itunes

[8] http://www.winamp.com

music organizers, users are not constrained to a fixed set of aspects by which to organize their music.

Aspects in our formalism form trees. The Nemoz taxonomy browser, depicted in figure 2, shows these hierarchical structures in a concise manner. This representation emphasizes the hierarchical structure of a Nemoz Knowledge Base. Users accustomed to organizing their media files in a file system might find this view most accessible.

In many situations, the dependence on a "fat client" is an obstacle to quick adoption. The immense popularity of recent "Web 2.0" media organizers like flickr hints to this fact. The Nemoz web interface plugin (see figure 3) embeds a web server to provide a simple way to browse and access items, categories and aspects of remote Nemoz users, without the need to install any software locally. Furthermore, it serves as a vehicle to explore user interface concepts for web-based multimedia organizers, following a "Representational State Transfer" design.

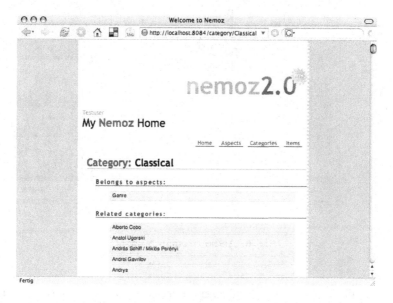

Fig. 3. Web interface

All popular "Web 2.0" media organizers feature an intuitive, fluid, incremental way of knowledge elicitation, afforded by their tag-based approach. In contrast to many heavier-weight methods of knowledge elicitation, tagging exhibits a high level of concreteness, which is an important trait of friendly user interfaces [25]. The supportedness property of our conceptual model naturally implies concreteness: The user is never forced to create empty categories or aspects that have to be filled with concrete content later. Instead, she structures her media content in a bottom-up manner, leading to concrete and meaningful categories and aspects.

<table>
<tr><th>name</th><th>Album</th></tr>
</table>

1:Black Crow	is contemplative	ther Roon
Jungle Fiction	[C:Überjam]
Nightingale	[C:Come Away With Me]
T	[C:Serenity (Disc 1)]

Fig. 4. Tagging items   **Fig. 5.** Tagging categories

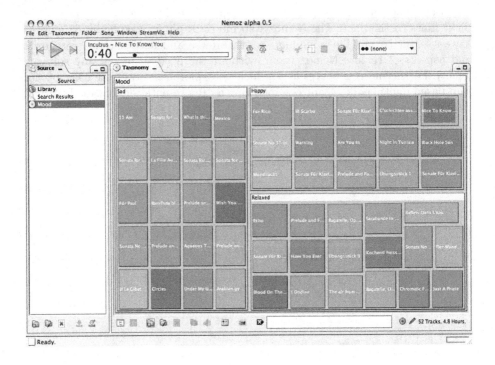

Fig. 6. Audio treemap browser

Almost all of our prototypical user interface variants support *"tag as you type"*, a simple way of tagging (and "untagging") the current selection of items or categories by directly typing a short english *tag phrase*. For example, the user might decide to annotate the song item "Black Crow" with the category tag "contemplative". To do so, she simply selects this song in a browser and types the phrase "is contemplative" (figure 4). While typing, the system might suggest completions, based on her own tags and other user's tags that where used to annotate similar songs. To associate the new category "contemplative" with the aspect "Mood", she simply selects this category in a browser and types "is a Mood" (figure 5). Note the article "a", by which the system distinguishes between adding to a super-category and adding to an aspect. Typing the tag phrase "is Mood" would have tagged the category "contemplative" with a new (super-)category "Mood". "Untagging" objects is achieved by negating a tag phrase (as in the tag phrase "is not contemplative") or by deleting the respective category or aspect.

The availability of content-based features allows a variety of visualizations that may be used to facilitate browsing of large media collections (for some recent examples, see [26] or [27]). We offer some simple visualizations in our Nemoz prototype, while more elaborate ones could be easily added as plugins. For example, the treemap [28] browser shown in figure 6 supports the user in structuring yet untagged music collections by coloring each song item according to its overall timbre.

6 Discussion and Conclusion

In this paper we have introduced aspects as a means to group tags. In particular, to each aspect, there is a hierarchy of tags. This allows to handle several hierarchies for one user. Whereas personalization approaches identify one user with one aspect, we take into account that the same user plays different roles depending on occasions. It also enables us to provide better services to users who organize their multimedia data. Retrieving, browsing, filtering becomes easy and accommodated to the user's personal aspects. Beyond enhanced human computer interfaces, the representation also allows more intelligent services. Automatic tagging using machine learning techniques for classification and unsupervised tagging using collaborative structuring reduces the burden of tagging.

The concepts and algorithms for aspect-based tagging are general, independent of the particular media which are to be structured. We have exemplified our approach by the Nemoz system which organizes music collections. Music collections are particularly hard to handle. For the user, a song must be listened to before she can tag it. In contrast, texts can more easily be skimmed through. For computation, music is given in a representation which must be converted into features. In contrast, texts already carry their primary ingredients of features, namely words. We have shown, how to deal with music by using a combination of feature extraction, feature transfer and reuse of tags. We denote the corresponding methods as "collaborative classification" and "collaborative structuring" respectively. They represent a new paradigm to web mining, that does not assume a global view on the resources on the Web, but a large number of partially related views among which we can selectively transfer information to optimize the system as a whole by providing intelligent assistance. We believe that this paradigm and corresponding approaches open the floor for new ways of user collaboration and better services for users.

References

1. Shardanand, U., Maes, P.: Social information filtering: Algorithms for automating "word of mouth". In: Proceedings of ACM CHI'95 Conference on Human Factors in Computing Systems, vol. 1, pp. 210–217. ACM Press, New York (1995)

2. Konstan, J.A., Miller, B.N., Maltz, D., Herlocker, J.L., Gordon, L.R., Riedl, J.: GroupLens: Applying collaborative filtering to Usenet news. Communications of the ACM 40(3), 77–87 (1997)

3. Linden, G., Smith, B., York, J.: Amazon.com recommendations: item-to-item collaborative filtering. Internet Computing, IEEE 7(1), 76–80 (2003)

4. Haase, P., Ehrig, M., Hotho, A., Schnizler, B.: Personalized information access in a bibliographic peer-to-peer system. In: Stuckenschmidt, H., Staab, S. (eds.) Semantic Web and Peer-to-Peer, pp. 141–156. Springer, Heidelberg (2005)

5. Schmitz, C., Hotho, A., Jschke, R., Stumme, G.: Mining association rules in folksonomies. In: Batagelj, V., Bock, H.-H., Ferligoj, A., iberna, A. (eds.) Data Science and Classification. Studies in Classification, Data Analysis, and Knowledge Organization, Berlin, Heidelberg, pp. 261–270. Springer, Heidelberg (2006)

6. Fielding, R.T.: Architectural styles and the design of network-based software architectures. PhD thesis, University of California (2000)

7. Baader, F., Calvanese, D., McGuinness, D., Nardi, D., Patel-Schneider, P.: The Description Logic Handbook. Cambridge University Press, Cambridge (UK) (2003)

8. Guoyon, F., Dixon, S., Pampalk, E., Widmer, G.: Evaluating rhytmic descriptors for musical genre classification. In: Proceedings of the International AES Conference (2004)

9. Lidy, T., Rauber, A.: Evaluation of feature extractors and psycho-acoustic transformations for music genre classification. In: Proceedings of the International Conference on Music Information Retrieval, pp. 34–41 (2005)

10. Schedl, M., Pampalk, E., Widmer, G.: Intelligent structuring and exploration of digital music collections. e und i – Elektrotechnik und Informationstechnik 7/8 (2005)

11. Moerchen, F., Ultsch, A., Thies, M., Loehken, I., Noecker, M., Stamm, C., Efthymiou, N., Kuemmerer, M.: Musicminer: Visualizing perceptual distances of music as topograpical maps. Technical report, Dept. of Mathematics and Computer Science, University of Marburg, Germany (2004)

12. Stenzel, R., Kamps, T.: Improving content-based similarity measures by training a collaborative model. In: Proceedings of the International Conference on Music Information Retrieval (2005)

13. Pampalk, E., Widmer, G., Chan, A.: A new approach to hierarchical clustering and structuring of data with self-organizing maps. Intelligent Data Analysis 8(2) (2005)

14. Logan, B.: Content-based playlist generation: Exploratory experiments. In: Proceedings of the International Symposium on Music Information Retrieval (2002)

15. Guo, G., Li, S.Z.: Content-Based Audio Classification and Retrieval by Support Vector Machines. IEEE Transaction on Neural Networks 14(1), 209–215 (2003)

16. Tzanetakis, G.: Manipulation, Analysis and Retrieval Systems for Audio Signals. PhD thesis, Computer Science Department, Princeton University (2002)

17. Pohle, T., Pampalk, E., Widmer, G.: Evaluation of frequently used audio features for classification of music into perceptual categories. In: Proceedings of the Fourth International Workshop on Content-Based Multimedia Indexing (CBMI'05) (2005)

18. Mierswa, I., Morik, K.: Automatic feature extraction for classifying audio data. Machine Learning Journal 58, 127–149 (2005)

19. Zils, A., Pachet, F.: Automatic extraction of music descriptors from acoustic signals using eds. In: Proceedings of the 116th Convention of the AES (2004)

20. Baumann, S., Hummel, O.: Using cultural metadata for artist recommendations. In: Proceedings of the International Conference on WEB Delivering of Music (2003)

21. Thrun, S., O'Sullivan, J.: Discovering structure in multiple learning tasks: The TC algorithm. In: Proceedings of the International Conference on Machine Learning (1996)
22. Mierswa, I., Wurst, M.: Efficient feature construction by meta learning – guiding the search in meta hypothesis space. In: Proceedings of the International Conference on Machine Learning Workshop on Meta Learning (2005)
23. Cohn, D., Caruana, R., McCallum, A.: Semi-supervised clustering with user feedback. Technical Report TR2003-1892, Cornell University (2000)
24. Wurst, M., Morik, K., Mierswa, I.: Localized alternative cluster ensembles for collaborative structuring. In: Proceedings of the European Conference on Machine Learning (2006)
25. Smith, R.B., Maloney, J., Ungar, D.: The self-4.0 user interface: manifesting a system-wide vision of concreteness, uniformity, and flexibility. In: OOPSLA '95. Proceedings of the tenth annual conference on Object-oriented programming systems, languages, and applications, pp. 47–60. ACM Press, New York (1995)
26. Moerchen, F., Ultsch, A., Noecker, M., Stamm, C.: Databionic visualization of music collections according to perceptual distance. In: ISMIR, pp. 396–403 (2005)
27. Knees, P., Schedl, M., Pohle, T., Widmer, G.: An innovative three-dimensional user interface for exploring music collections enriched. In: MULTIMEDIA '06. Proceedings of the 14th annual ACM international conference on Multimedia, pp. 17–24. ACM Press, New York (2006)
28. Johnson, B.: Treeviz: treemap visualization of hierarchically structured information. In: CHI '92. Proceedings of the SIGCHI conference on Human factors in computing systems, pp. 369–370. ACM Press, New York (1992)

Contextual Recommendation

Sarabjot Singh Anand[1] and Bamshad Mobasher[2]

[1] Department of Computer Science, University of Warwick, Coventry CV4 7AL, UK
s.s.anand@warwick.ac.uk
[2] Center for Web Intelligence, School of Computer Science, Telecommunications and
Information Systems, DePaul University, Chicago, Illinois, USA
mobasher@cs.depaul.edu

Abstract. The role of context in our daily interaction with our environment has
been studied in psychology, linguistics, artificial intelligence, information retrieval, and more recently, in pervasive/ubiquitous computing. However, context
has been largely ignored in research into recommender systems specifically and
personalization in general. In this paper we describe how context can be brought
to bear on recommender systems. As a means for achieving this, we propose a
fundamental shift in terms of how we model a user within a recommendation
system: inspired by models of human memory developed in psychology, we distinguish between a user's short term and long term memories, define a recommendation process that uses these two memories, using context-based retrieval
cues to retrieve relevant preference information from long term memory and use
it in conjunction with the information stored in short term memory for generating
recommendations. We also describe implementations of recommender systems
and personalization solutions based on this framework and show how this results
in an increase in recommendation quality.

1 Introduction

The role of recommender systems in addressing the information overload problem
is well established with a number of commercially available recommender systems
providing benefits to both users and businesses. However, most currently available recommender systems still tend to use very simplistic user models to generate recommendations. For example, user-based collaborative filtering generally models the user as a
vector of item ratings and content based filtering methods tend to use models such as
the naïve Bayes and bag-of-words or feature vectors.

The user models also tend to be additive in nature. For example, in user-based collaborative filtering, as more ratings are provided by the user, they are simply added to the
existing set of ratings and all item ratings are used in discovering the active user's neighbourhood. Similarly, content based techniques tend to just update the bag-of-words or
probabilities as new items are rated. A partial exception to such an additive approach
is the work on the Adaptive Information Server (AIS) [1]. In AIS, Billsus and Pazzani, distinguished between long term and short term interests of a user. They used *tfidf*
scores for words appearing in the last 100 documents accessed by the user for modelling short term interests while long term interests were modelled using *tfidf* scores
for words appearing in all documents accessed by the user. Hence while a distinction

B. Berendt et al. (Eds.): WebMine 2006, LNAI 4737, pp. 142–160, 2007.

is made between long and short term interests, the long term interests are essentially additive and the size of the short term interests is arbitrary.

This additive approach to modelling the user simply ignores the notion of "situated actions" [2], that is, the fact that user's interact with systems within a particular "context" and ratings for items within one context may be completely different from the rating for the item within another context. It is therefore not surprising that stories of inappropriate recommendations abound, such as the male customer buying a pregnancy book from Amazon.com as a present, persistently receiving recommendations on pregnancy related topics [3].

More concretely, consider the example of a user who buys and rates books of contemporary fiction for himself (e.g., "Gravity's Rainbow"), work-related books on computer science topics (e.g., "Programming Python"), and books for his children (e.g., "Where's Waldo?"). It makes little sense to represent this user's "interest in books" in a single representation that aggregates all of these disparate works. Yet that is precisely what most recommender systems will do. An additive representation loses rather than gains predictive power as multiple contexts are combined. Preferences expressed in one context such as "children's books" will be of no predictive value when recommendations in a different context "computer books" are sought, and in fact, will act as distracters generating the false impression of similarity between users. The ideal contextual recommendation system would therefore be able to reliably label each user action with a context. Thus, neighbors with similar tastes in children's books would be used only when the "children's book" context is active and would be ignored otherwise.

While little agreement exists among researchers as to what constitutes context, the importance of context is undisputed. In psychology, a change in context during learning has been shown to have an impact on recall [4, 5], suggesting a key role played by context in structuring of and retrieval from human memory. Research into linguistics has shown that context plays the important role of a disambiguation function, that is, it reduces the possible interpretations of a message that exists in abstraction from its context [6].

In this paper we present a novel approach to incorporating user context within the recommendation process. We model the user based on human memory models proposed in psychology. Preference models for previous user interactions with the system are stored as memory objects within the user's Long Term Memory (LTM) while the preference model of the current user is stored in the user's Short Term Memory (STM). Contextual Cues generated from the data stored in STM are used to retrieve relevant objects from LTM which are then used to generate recommendations for the user.

For example, in the case of the hypothetical book buyer mentioned above, we might imagine three models stored in LTM: one for contemporary fiction M_1, one for children's books M_2, and another for computer books M_3. If the user were browsing the children's section of a book catalog, it would be appropriate to retrieve M_2. This preference information would then be combined with whatever information was currently being gathered from the user's interaction to form the basis for recommendation.

If recalling the appropriate context were simply a matter of identifying genres, as in this example, then there would be little complexity involved. However, the distinguishing features of a context may be considerably more subtle. A kindergarten teacher may buy children's books for her own classroom use as well as for her children, for example. One of our important assumptions is that context is not necessarily an observable feature of an interaction. Behavior is observed and this behavior is induced by an underlying context, but the context itself may not be directly observable. This assumption distinguishes our approach from previous work which defines context as a fixed set of attributes such as location, time or identities of nearby individuals or objects, as is commonly done in ubiquitous computing [7].

The key contributions of this paper to recommender systems research are:

- a new approach to modeling users, based on research in psychology, consisting of short and long term memories (STM and LTM, respectively) that incorporates the notion of user context;
- the definition of a process, based on contextual retrieval from LTM, to generate recommendations of a higher quality than those generated using traditional user models;
- a classification of contextual retrieval cues and how they can be used in the recommendation process;
- a description and evaluation, using real data from an e-tailer, of approaches to collaborative recommendation that use the new user model;

2   Modeling Context

Dourish distinguishes between two views of context: the representational view and the interactional view. He suggests that the representational view, dominant in ubiquitous computing, makes four key assumptions [8]: context is a form of information, it is delineable, stable and separable from the activity. What this means is that context is information that can described using a set of "appropriate" attributes that can be observed, hence collected. Furthermore, these attributes do not change and are clearly distinguishable from features describing the underlying activity undertaken by the user within the context.

In the representational view, incorporating context within context-aware applications is generally viewed as a process consisting of a number of technological steps such as sensor fusion, feature extraction, classification and labeling [9] leading to context recognition and prediction. Once identified, context has been used to label and store user interactions for recall in the future [10] or for dynamically adapting the system's interface to the user [11].

Lieberman and Selker [12] define context as "everything that affects the computation except its explicit input and output". This definition includes the state of the user, state of the physical environment, state of the computational environment and history of user-computer-environment interaction. A less computationally focused definition of context is provided by Dey [13], as "any information that can be used to characterize the situation of an entity". Dey further elaborates on the term, situation, as "typically the location, identity and state of people, groups and computational and physical objects". Schilit and Theimer [7] add to this representational view of context by including "lighting, noise level, network connectivity, communication costs, communication bandwidth and social situation".

One of the key issues with this view is that the definition of context is limited to those elements that are observable. Hence in the case of interactions on the Web, we may consider the time of day and user agent (providing information about the device used) as being key contextual attributes, assuming that we are not adding any sensors to monitor the user's environment. Additionally, some metrics can be derived from the user behavior such as the amount of scrolling, the speed of browsing etc.

In the absence of any precise theory of what defines context for web interactions, it is difficult to restrict the set of allowable contextual features. Indeed, it is hard to exclude any feature a priori as it is always possible to construct an example in which that feature is key. For example, should there be contextual features corresponding to the user's choice of clothing? It might not seem necessary, but if a restaurant recommendation is needed, the recommender may wish to exclude establishments whose dress code is too formal for the user's preferred attire. Researchers in ubiquitous computing tend to get stuck in this trap: building models of context that consist of many features, the majority of which are relevant only in limited circumstances and to a limited set of users, because it is hard to rule any out. We suggest that it is the dominance of the representational view of context and its attendant complexities that is responsible for the reluctance of recommender systems researchers to build context-enhanced systems.

The interactional view of context takes a different stance on each of the four assumptions made by the representational view. In the interactional view, Dourish suggests that contextuality is a relational property, i.e. some information may or may not be relevant to some activity. He also proposes that the scope of contextual features is defined dynamically, and occasioned rather than static. Finally, rather than assuming that context defines the situation within which an activity occurs, Dourish suggests a cyclical relationship[1] between context and activity, where the activity gives rise to context[1]. The key distinction between these two viewpoints on context is that while the former is concerned with what context is and how it can be represented within an application, the latter is concerned with "achieving and maintaining a mutual understanding of the context for their (user and system) actions", no matter how the context is actually defined.

We adopt the interactional view as the basis for our approach to modeling context. As noted above, we make the assumption that the observed user behavior is induced by an underlying context, but that the context itself is not necessarily observable. This assumption frees us from limiting our definition of context to a fixed set of attributes, nor do we assume the existence of a concept hierarchy [14] or pieces of text [15] as is often assumed in information retrieval applications. In fact, we suggest that the precise nature or representation of the context is not as important as recognizing the existence of, and accurately predicting the user context from, a set of known or derived contextual states.

In general, a user interacts with the web by requesting resources available on web servers. This interaction may take the form of navigating through a sequence of pages using hyperlinks; providing parameters to an application, such as a search engine query, or providing implicit or explicit indications of interest, such as ratings, for various items.

---

[1] Note that Lieberman and Selker [12] also allude to this cyclical relationship when they state that the behavior of an application, provided with explicit inputs as well as context (implicit inputs), can be altered not only to affect the explicit output but may also the context itself.

Standard methods now exist for pre-processing of clickstream data to associate a request with a user and partition a user's activity into visits or sessions [16]. An interesting research problem is that of automatically segmenting user activities based on the context within which these activities occur. Hence a new interaction begins when the user makes a transition from one contextual state to another, and thus, a user's context within an interaction is static and may span multiple visits or be part of a single visit. In the example applications of Section 4 we use a simple heuristic to achieve this segmentation, that a single user interaction constitutes the activity of the user during one session. Evaluation of our approach to recommendation, presented in Section 4.3, suggest that the benefits accrued by contextualizing recommendation generation by far outweighs the inaccuracy in segmentation due to the use of this heuristic. Approaches based on Hidden Markov Models may provide more accurate segmentation of the user's page view sequence, where the hidden states are the user contexts while the observed states are the items rated by the user.

Contrary to the above simplified interaction model, systems such as Netflix[2] and MovieLens[3] attempt to get the user to build their profile by providing ratings for as many movies as possible on registration. Building a contextual model of the user in this case requires techniques for discovering contextual states and segmenting user item ratings into a set of interactions. How this is achieved, remains as open research question. Possible approaches to solving this problem may include the use of generative probabilistic models that assume the observed behavior is generated by an underlying set of hidden factors (representing the user's context). This is the approach used by Xin et al. to model user interactions based on the underlying "tasks" being performed [17]. Hence, in general we use the term interaction to mean the "logical" interactions of the user with their environment which may span one or more physical interactions.

We propose that context can be modeled as a stochastic process that is in one of the d states defined by the set $C = \{c_1, c_2, ..., c_d\}$, representing the distinct contexts within which a user u_k interacts with the system.

Clearly an important design consideration is the value of d, i.e. the number of distinct contexts within which a user may interact with the recommendation system. Deciding on the value of d is similar to the long standing problem in clustering, of choosing the number of clusters within a given dataset. Various methods have been proposed as a solution to this problem based on maximum likelihood and cross validation [18]. Thus we could choose d, in a similar way, so as to maximize the likelihood of the observed user behavior.

3   Recommendation Framework

Let us assume that we have a set of m users, $U = \{u_k : 1 \leq k \leq m\}$, and a set of n items, $I = \{i_j : 1 \leq j \leq n\}$. Let $u_a \in U$, referred to as the *active user*, represent the user whose navigation through I needs to be personalized. In previous interactions, u_a will have either explicitly or implicitly rated a set of items $I_a \subset I$. Typically, these systems assume that the user u_a rates items in I using a rating function r_a, defined as

---

[2] www.netflix.com

[3] www.movielens.umn.edu

$r_a : I \rightarrow [0, M]$, where M is some maximum rating value defined by the system, that reflects the user's level of interest in a particular item.

We refer to the set $Q_a = I - I_a$ as the *candidate item set* for the user u_a. The goal of the recommendation engine is to select a set of items, $R_a \subseteq Q_a$ consisting of items of interest to the active user. This is achieved by approximating r_a from the ratings in I_a and any other data made available to the recommender system, typically, an item knowledge base and ratings by other users in U.

Various classifications of recommender systems have been proposed in literature [19, 20, 21]. Among these the most commonly advocated approaches are Collaborative and Content-based Filtering. Both techniques require the building of a user model. Collaborative Filtering, traditionally a memory based approach to recommendation, represents each user as a vector of item ratings. Using similarity metrics, the algorithm first generates the active user's neighborhood, consisting of k most similar users to the active user and then recommends those items to the active user that have been highly rated by his neighbors. Content based filtering on the other hand builds a model of the user's likes and dislikes in terms of the content descriptors of the items.

In either case, we suggest that the context within which the items are rated by users should form part of the user model. This allows the user to potentially rate the same item multiple times within different contexts. For example, rating a particular movie highly when the context is "watching a movie with your partner" as opposed to "with your boss". Hence, rather than a single rating function, r_k as described previously, we propose the existence of d rating functions, r_{k_i}, one for each context. Thus item ratings are specific to the user and the context within which the item was rated.

In the following sections we describe our proposed framework for a contextual user model and how such a model can be used to generate recommendations.

3.1 Contextual User Models

Our model is inspired by Atkinson and Shriffin's model of human memory [22], which is still the basis of our current understanding of the structure of human memory. This model consists of three structural components of human memory: the sensory register, the short term store, and the long term store. According to the model, when stimulus is presented, it is immediately registered within the sensory register. A scan of the information within the sensory register leads to a search of the long term store for relevant information, leading to the transfer of information to the short term store from the sensory register and long term store. The data within the sensory register and short term store decay with time, generally within a very short time period, whereas the long term store is more permanent. In addition to these three structural components, the model also identifies control processes such as transfer between short term and long term stores, storage, search and retrieval within short and long term storage.

Raaijmakers and Shriffin proposed a probabilistic theory of how search is conducted within the LTM [23]. In their model, LTM is considered to be a highly interconnected network of memory objects. In addition to memory objects a retrieval structure is defined based on probe cues and their strengths of association with memory objects. Contextual cues are identified as an important part of this retrieval process. When faced with a particular question, a retrieval plan is generated which in turn is used to assemble a set of probe cues. Memory objects are retrieved, based on the degree to which they are associated to the cues in comparison to other objects, and evaluated possibly leading to a successful retrieval or a refinement of the retrieval plan for the next iteration of search within LTM.

This model fits in quite well with our needs. From the perspective of a recommender system, a user interacts with the system through implicit and explicit input. These inputs constitute the active interaction of the user and can be thought of as being stored in the short term store (the sensory register being implicit in this interaction). User preferences implicit in previous interactions of the user are stored within the long term store as memory objects, as described below. Some of these may be relevant to the active interaction as they took place within the same context as the active interaction. Hence these memory objects must be retrieved and transferred to the short term store for processing, i.e. recommendation generation. The cues for this retrieval are generated from data collected during the active interaction. Finally, the user preferences implicit within data collected during the active interaction will be extracted and transferred to the long term store for use in future interactions.

Hence, in modeling the user, u_k, we distinguish between two types of memories: Short Term Memory (STM) and Long Term memory (LTM). The user's Long Term Memory is modeled as a set of memory objects. Each memory object is a two-tuples $\langle c_{k_i}, r_{k_i} \rangle$ where, r_{k_i} is a user preference model derived from previous interactions and c_{k_i} is the context within which the interaction modeled by r_{k_i} took place. Contextual retrieval cues defined in Section 3.3 are probabilistically associated with these memory objects.

At a basic level, user preferences can be model-based or memory-based [20]. In memory based approaches, the preference model is generally a vector of item ratings or a vector over item attributes. In addition to user ratings of items within I, in the presence of an item ontology, the user preference models can take the form of an instance of an ontology with weights, associated with edges, representing the significance of relationships between objects and their attributes. Model-based approaches such as naïve Bayes, clustering, association rules and sequence patterns have also been used for modeling user preferences and have shown to provide scalability for recommendation generation. In the contextual user model proposed here we are not prescriptive about the type of modeling approach used to capture user preferences, as long as the model is able to meet the requirements of the key processes laid out in the next section. Specifically, preference models that are incremental and can be combined without the need to rebuild them from the underlying data are preferable for computational efficiency (in particular refer to the processes related to merging of preference models with active interaction ratings and updating preference models within LTM).

The user's short term memory is the working memory, where user preference data provided by the active user within the active interaction is stored, retrieval cues are generated and any preference models of the user transferred from the user's LTM using the retrieval cues are stored.

3.2 Recommendation Generation

The task of generating recommendations can be summarized as follows. Explicit or implicit ratings for items from the active interaction are stored in the STM. Contextual cues are then derived from this data and used to retrieve some preference models from the user's LTM that are deemed to belong to the same context as the active interaction. These are merged with the ratings stored in the STM and are then used to predict ratings for items not currently rated by the user.

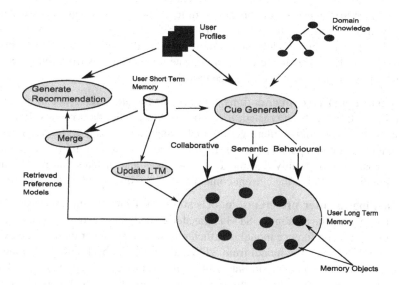

Fig. 1. The Contextual Recommendation Process

Thus the key processes involved within this framework are the following.

(1) Generation of contextual cues from behavioral data stored in the STM. Cues extracted depend on various factors such as the extent of the user history available, the availability of an item knowledge base and the amount of data available on the active user interaction. For example, depending on the application, these cues may be based on users' ratings of items, the amount of time spent on a page view, the textual features of documents viewed, semantic properties of objects of interest as might be available through a domain ontology, keyword queries used in search, or other implicit or explicit measures of interest. We further discuss several types of contextual cues in Section 3.3.

(2) Retrieval of relevant memory objects from LTM. This process requires the ability to create local abstractions, r_a representing the preference model of the active user u_a's interaction. Depending on the representation of memory objects stored in LTM, these local models may be represented as vectors of item ratings; as weights associated with concepts or textual features of documents; or, in the presence of a domain ontology, "ontological user profiles" which are instances of an ontology with weights associated with objects, attributes, or relations [24]. The task of identifying relevant memory objects from the LTM is then reduced to computing similarities or associations between the local models for the interaction and the preference models stored in the LTM.

(3) Merging the retrieved preference models with data in the STM. Once the relevant memory objects have been retrieved from the LTM, they must be incorporated into the preference model associated with the current interaction. Generally, this task involves the aggregation of multiple preference models (e.g., rating functions) into a single representation. For example, in a typical collaborative filtering application, the retrieved memory objects would provide additional ratings on various items which, together with the ratings in the STM form the preference model of the interaction. If the retrieved preference models have a more structured or ontological representation, the merging task will involve creating aggregate representations from multiple object instances [25] or an ensemble preference model.

(4) Generation of recommendations. Once the aggregated preference model for the active interaction has been created, it can be used to predict item ratings. In the case of user-based collaborative filtering, only that rating function is used for neighborhood formulation that corresponds to the user's predicted context[4]. Similarly, in the case of content based filtering, only that model of user likes and dislikes that represents the user's preferences within the current context is used to compute recommendations.

(5) Updating the user preference models stored in LTM. At the conclusion of the active interaction, the data in the STM is transferred and integrated into the LTM. If the active interaction preference model is sufficiently similar to one or more of the existing contexts in LTM, then the model from STM can be merged back into those long-term memory objects in a manner similar to the merging process described above. On the other hand, if the user's activity represents a new context, the current model is added as a new memory object, effectively reorganizing the retrieval structure within the user's LTM.

3.3 Contextual Retrieval Cues

In Section 3.2 we described how STM and LTM interact through the use of contextual retrieval cues to generate recommendations.

The process of selecting a context from the LTM can be understood in Bayesian terms. Let L_i be a memory object from LTM corresponding to a context. We are interested in selecting the most likely context given the current state of the user interaction, that is, the L_i such that $p(L_i|STM)$ is maximized. The contextual aspects of the interaction are summarized in contextual cues CC_j extracted from the STM. The probability can then be calculated as follows:

$$p(L_i|STM) = \frac{1}{p(STM)} \sum_j p(L_i|CC_j)p(STM|CC_j)p(CC_j)$$

The value $p(L_i|CC_j)$ corresponds to the probabilistic association between a contextual cue CC_j and a particular context from the LTM L_i. The value $p(STM|CC_j)$ is the probability of a given set of user observations given the validity of a particular cue

---

[4] The identification of context does not imply the prediction of a definitive context of the user. Rather it is the definition of $P(C|STM)$, where STM represents the current contents of the user's STM.

CC_j. The calculation of this value would be highly dependent on the particular type of cue being used. Consider a simple semantic cue, such as book genre. In this case, the value $p(STM|CC_j)$ could be calculated as the fraction of positive ratings in the STM associated with books of the given genre. The prior probabilities $p(STM)$ and $p(CC_j)$ would be estimated from historical data.

We now identify three different types of cues that can be generated from data stored in STM and discuss how these cues are generated. The key requirement that these cues must meet is that they must reflect different user contexts. We provide empirical evidence of their value within the recommendation process in Section 4. The first two categories of cues represent different methods for computing similarity between preference models. The third category (Behavioral cues) uses behavioral aspects of the user during the interaction rather than the actual ratings of items as the basis for computing similarity.

(1) Collaborative Cues: Collaborative cues represent items as m-dimensional vectors consisting of ratings for the item by the m users of the system. Memory objects from LTM with preference models that have a similarity greater than a particular threshold are retrieved and placed within the active user's STM for use in the recommendation generation process during the active interaction.

(2) Semantic Cues: Semantic cues are similar to collaborative cues in that they measure similarity of the user preference model from the active interaction with those stored in the user's LTM and retrieve those interactions from LTM that have a similarity, greater than a pre-defined threshold, with the active ratings. However these cues assume the existence of an item knowledge base and use item semantics to compute similarity between items. If items of interest are text-based documents, then textual features and weights can be obtained using methods such as the standard *tfidf* approach commonly used in information retrieval.

(3) Behavioral Cues: Various metrics such as velocity, search-to-browse ratio and amount of scrolling may be used as metrics to describe user behavior on a web site. Similarity between these metrics computed for the active interaction and previous interactions of the user are used as the basis for retrieving past interactions from the user's LTM. An alternative approach, when an item ontology is available, is to extract latent factors that drive user choice, for example, impact values extracted using Kullback-Leibler's Information Divergence [24] and use these as the basis for describing user behavior (see Section 4).

During recommendation generation, these cues need not be used in isolation. In Section 4.2 we discuss how a hybrid cue that combines instances of semantic and behavioral cues can be used.

3.4 Effect of Contextualization on Sparsity of the Rating Matrix

Given that the rating matrix used in collaborative filtering is already known to be sparse when not considering user context, an obvious concern is whether contextualizing the recommendation process would further degrade performance of the recommender system due to even greater sparsity being introduced as each user is being further split along a context dimension. For example, consider the rating matrix in Table 1. User 1 may have provided the ratings through two previous interactions, resulting in the user model shown in Table 2. Note that the user model in Table 2 also shows that the user has rated two new items, namely, "The Sixth Sense" and "Unbreakable", in the current interaction, preference data that is stored in user 1's STM.

Table 1. Rating Matrix

User	The Sixth Sense	Unbreakable	Star Wars	Cliff Hanger	Armageddon	Bandits	Die Hard	The Terminator	X-Men	Gladiator
1			4	4	4	5	4			
2							3	5	4	
3	3		3				4			
4		2	4		5				4	3
5	3			3			4	4	5	

Table 2. User Profile for User 1

Memory	Interaction#	The Sixth Sense	Unbreakable	Star Wars	Cliff Hanger	Armageddon	Bandits	Die Hard
STM		5	5					
LTM	1			4	4			
	2					4	5	4

In [24], Anand et al. showed how using a metric such as the Generalized Cosine Max, that utilized item similarity within the calculation of user similarity can greatly reduce the negative effects of sparsity on recommendation accuracy.

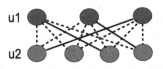

Fig. 2. Calculating User Similarity

Consider two users u_a and u_b. We can model the two users as a bipartite graph (see Figure 2), $G = \langle V, E \rangle$, where ratings for items provided by each user represents the two disjoint sets of vertices and an edge between pairs of nodes $\langle i_{a_j}, i_{b_f} \rangle$, where $i_{a_j} \in I_a$ and $i_{b_f} \in I_b$ has a weight defined by a similarity metric $sim(.,.)$ defined on the set of item pairs. The calculation of similarity between the users can be viewed as an instance of the *Assignment Problem* that aims to find a maximum weight matching[5] in a

---

[5] A matching is a subset of edges of a weighted bipartite graph such that no two edges have a node in common. In Figure 2, the solid edges show one such matching. Note that the weights are not shown in the figure.

weighted bipartite graph. The Hungarian algorithm [26] is the best known algorithm for solving the assignment problem in $O(n^3)$ where $n = max(|I_a|, |I_b|)$. The Generalized Cosine Max (GCM) metric [24], implements such a strategy for measuring similarity between users and is defined as

$$u_a \cdot u_b = \sum_{(i_j, i_f) \in S} r_a(i_j) \times r_b(i_f) \times sim(i_j, i_f) \tag{1}$$

where, r_a is the user preference model of the user u_a and r_b, is the user preference model for u_b and $S \subset E$ is the matching of the bipartite graph as defined above[6].

For example, consider the rating matrix in Table 1. The distance between users 1 and 2 is calculated using the item similarities matrix shown in Table 3. Here S = {⟨Cliff Hanger, Gladiator⟩, ⟨Die Hard, The Terminator⟩, ⟨Armageddon, X-Men⟩} and the similarity between the users is calculated to be 0.32 despite the fact that the users have no item ratings in common.

Table 3. Item Similarity

Item	The Terminator	X-Men	Gladiator
Star Wars	0.37	0	0
Cliff Hanger	0	0.34	0.42
Armagageddon	0.37	0.37	0
Bandits	0	0.34	0.42
Die Hard	0.44	0.43	0

The evaluation of our contextual recommendation approach (see Section 4.3) shows that despite the increase in sparsity, the contextual recommendation approaches actually improve recall by approximately 25%. Hence the benefits of contextualization appear to outweigh the effects of greater sparsity within the rating matrix.

4 Example Application: Contextual Collaborative Filtering

Consider a user, u_a, say User 1 in the example in Tables 1 and 2, that requires movie recommendations. Rather than generating a crisp partition of previous interactions into d classes representing the d contextual states, and generating a single memory object for each state that aggregates the user preference models from the interactions, we use a memory based approach to representing the user preference model for the user within a context. Hence each previous interaction is stored within the user's LTM as a single memory object.

---

[6] Note that the original definition of the Generalized Cosine Max [24] used a greedy algorithm for selecting the edges in S rather than the Hungarian algorithm.

4.1   Item-Based Collaborative Context Discovery

In this application we assume that the user preference model is represented by an n-dimensional rating vector. Collaborative cues are modeled by representing items as m-dimensional vectors consisting of ratings for the item by the m users of the system. Hence similarity between the STM and L_i's can be calculated using the GCM metric described in Section 3.4. As each item i_j is defined as an m-dimensional vector, the item similarity function, $sim(i_j, i_f)$, for purposes of this evaluation, is calculated as the cosine similarity between items i_j and i_f $(i_j \cdot i_f)$.

To deal with the issue of *sufficient similarity* only those memory objects are retrieved from the user's LTM that have a similarity to the active preference model greater than a predefined user similarity threshold. As the user preference models retrieved from LTM are rating vectors, merging them with ratings from the active interaction is trivial. The aggregate user preference model is then used to define a neighborhood for the active user. The neighborhood consists of the k most similar interactions of the system with user's other than the active user. Finally item ratings are generated from the neighborhood using the standard weighted sum approach used in traditional collaborative filtering. The final step of updating the active user's LTM consists of simply the creation of a new memory object that stores the user preference model consisting of those item ratings that were provided by the user within the active interaction.

Using the GCM for the example in Tables 1 and 2 we get similarity values, for STM with LTM1 and LTM2, of 0.56 and 0.49 respectively. Assuming a similarity threshold of 0.5, LTM1 will be retrieved and merged with STM prior to the calculation of the user's neighborhood.

4.2   Semantics Based Collaborative Context Discovery

When an item ontology is available, it can be used in two ways to generate cues (Semantic and Behavioral, respectively) for retrieving previous user preference models from the user's LTM:

- To calculate similarity between user preference models based on semantics rather than item ratings as described in Section 4.1. In this case LTM2 would be more similar to the user's STM, given that "Bruce Willis" has acted in all of the films in the STM and LTM2.
- To discover impact values, i.e. latent factors driving item ratings by the user within the interaction. The ratings in the user's STM may be used to discover whether the user's preference for these movies is based on an interest in movies acted in by "Bruce Willis" or movies directed by "M. Night Shyamalan". If the latter is true, then a better interaction to retrieve from the user's LTM may be one that is also based on the user's interest in a director.

We now describe a particular implementation of our user model. Given the availability of an item knowledge base (instances of an ontology similar to the example shown in Figure 3), the user preference model is now in the form of an ontological profile [24] consisting of a set of instances of the item ontology and a set of weights associated with the edges of the ontology. The ontological profile is generated from the rating data

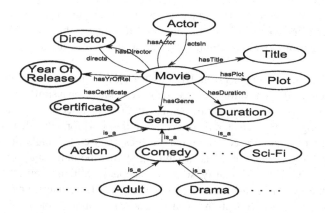

Fig. 3. Example Movie Ontology

collected from a user within a single interaction and the item ontology. Specifically, the Kullback-Leibler's Information Divergence metric was used to generate the weights (impact values) associated with the edges of the ontology. The impact weights are defined within the interval $[0, 1]$ where a larger weights associated with the edge suggests that the user ratings of movies within an interaction were more strongly influenced by that particular concept.

The impact of a concept Q_l on the item selection by a user can be viewed as the divergence of the *observed* distribution, f, of instances of Q_l within an interaction, v, from an *expected* distribution, g. The expected distribution is what would be observed if the concept indeed had no effect of the users item selection i.e. it is the same as a user that randomly selects items with instance of Q_l that conform to some background distribution. An appropriate background distribution in the case of recommender systems is one that reflects the distribution of instance of Q_l within the item knowledge base.

This divergence is measured using the Kullback-Leibler information divergence metric defined as [27]

$$imp = \sum_{x \in D_l} f(x) \log \frac{f(x)}{g(x)} \qquad (2)$$

where, the sum over x indicates that x is a random variable describing a possible occurrence of an instance of the concept in a visit and summed over all possible instances of the concept.

The advantage of using $g(x)$ as defined above, rather than assuming a uniform distribution, is that the impact measure now incorporates the prior probability of the instances of C_l. Let us consider an example to illustrate this point. Two visitors to a movie web site show a preference for movies with the year of release $\{1930, 2000\}$ and $\{1999, 2001\}$ respectively, with equal probability. However, if it is known that only 10 movies out of the 10000 movies held in the retailers database were released in the year 1930 whereas 1999, 2000 and 2001 saw an equal number of releases, numbering 1000, using this information to define $g(x)$ will result in a higher impact value for *year of release* for the first visitor (5.64) as compared to that for the second visitor (2.32), whereas using a uniform distribution as $g(x)$ would return the same value for both visitors.

Retrieval of memory objects from LTM uses a hybrid cue consisting of a Behavioral cue in the form of impact values and a Semantic cue based on a similarity function that computes item similarity based on the similarity of their descriptions in accordance to the item ontology. The interested reader is referred to [24] for a detailed description of the item similarity metric used. The impact values are used to weight the similarity of concepts used to describe the items, when aggregating the individual concept similarities into the overall item similarity, so that concepts with higher impact have a larger bearing on the overall item similarity than concepts with smaller impact values. The GCM metric, as defined in Section 4.1, uses this item similarity metric to compute the similarity between user preference models.

The process of merging retrieved user preference models with the user preference model of the active interaction requires the recomputation of the impact values as the observed distribution of instances of the concepts is likely to have changed. As in the case of the item-based context discovery system, once the aggregated user preference model has been generated, the active user's neighborhood is defined and used to generate item predictions. Updating the user's LTM also simply requires the creation of an ontological profile using ratings collected during the active interaction and storing it as a new memory object.

4.3   Evaluation

Table 4 shows the results of generating recommendations using web log files from a movie retailer. Four algorithms were evaluated. *RandomNeighbour* is a baseline algorithm that randomly assigns users to the active user's neighborhood. Any algorithm using an appropriate similarity metric should improve on such a random neighborhood selection algorithm. *Traditional CF* bases the neighborhood of the active user on all items rated by the active user in all previous interactions while the *ContextualRecommender* implements the approach to recommendation described in Section 4.1. The *ContextualSemanticRecommender* algorithm implements the approach to recommendation briefly described in Section 4.2, using the Semantic cue as the basis for retrieval from LTM. For each of the algorithms the neighborhood size used was 50.

Of all the visits to the web server over a three month period, for this evaluation, we selected visits that had a minimum of 10 rated items and a maximum of 50 rated items. The total number of unique visitors and visits meeting this criteria were 31,223 and 54,964 respectively consisting of 923,987 ratings. Of these visits, 8572 visits were randomly selected as the test data, restricting the choice of visits to only those visitors with at least two visits so that a minimum of one LTM object exists for each visitor within the test data. Figure 4 shows the number of objects in LTM for each of the visitors in the test data. Of the visits selected as test data, five items were randomly selected from each visit and hidden from the recommender system. The remaining data was used to model the short term memory of the active user. Items within the data set

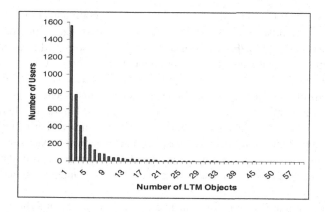

Fig. 4. Number of LTM Objects per Visitor in the Test Data

were movie, actor or director pages rated implicitly by the user defined as the log of linger time of the user on the page. The total number of items exceeded 140,000.

As can be see from Table 4 the contextual recommender improves on precision and recall compared with the traditional approach to user-based collaborative filtering. Also, the semantic cues seem to (slightly) outperform the collaborative cues. While our model improves on traditional collaborative filtering in both, precision and recall, the improvement in recall is clearly the more significant of the two, ranging between 25-30%.

Table 4. Evaluation Results (Large Short Term Memory)

Algorithm	Precision	Recall	F1
RandomNeighbour	80.4%	1.7%	0.033
Traditional CF	80.45%	8.22%	0.149
ContextualRecommender	83.8%	10.38%	0.184
ContextualSemanticRecommender	84.3%	10.81%	0.191

5 Related Work

Adomavicius et al. [28] proposed a recommendation approach that incorporates contextual information using a multidimensional approach. Their proposal strongly aligns itself with the representational view of context. As a result, the user must provide explicit contextual data, such as the time, place and companion, in addition to item ratings. It is unclear as to whether the additional burden placed on the user will result in a reduction in usage of the system based on privacy concerns [29] or indeed the inertia associated with providing data that is not seen as central to the task at hand [30]. The inaccuracy in demographic data collected through various registration processes on the web would suggest a similar fate for such contextual data collection.

Hayes and Cunningham [31] also identified the issue of additive user profiles as being a weakness of collaborative filtering for their application of recommender systems to personalize a web-based music service. The unit of recommendation was a playlist, a compilation of ten music tracks built up by a user. The recommendation generation process, modeled on the MAC/FAC model of retrieval [32], consisted of an initial collaborative filtering step that generated a recommendation list that was then further refined by a content based filtering step that used the genre and artist tags associated with the playlist. The user context was defined by the current playlist. Hence those playlists that are in the recommendation list generated by the initial collaborative filtering step and were similar to the current playlist, with regard to their genre and artist tags, were promoted to the top of the recommendation list. In their system, context is not persisted and used for future use when generating recommendations, hence making the user profile for all user other than the active user still additive. Also, context is limited to semantic cues only.

In AIS, Billsus and Pazzani, distinguished between long term and short term interests of a user [1]. They used *tfidf* scores for words appearing in the last 100 documents accessed by the user for modeling short term interests while long term interests were modeled using *tfidf* scores for words appearing in all documents accessed by the user. Our user model, while consisting of a long and short term memory is very different from that introduced by Billsus and Pazzani. Firstly, the long term memory in our model incorporates the notion of context and hence is not simply a single vector describing the user's long term interests as is the case in AIS. Incorporating context within long term memory allows us to go beyond the notion of concept drift over time to include cyclic interests of users. Secondly, in AIS, long term memory is only invoked if the short term memory is not able to suitably classify a news story as being of interest or not to the user. In our case, portions of long term memory deemed to have originated in previous user actions within a similar context to the current interaction always augment the short term memory.

Gasparetti and Micarelli proposed a user profile based on memory retrieval theory [33], specifically on Associative Memory. Cues were generated using the notion of information scent [34], however, the notion of context, while mentioned in the paper was not evaluated. Further, they only described how cues can be generated from web documents, hence only catering to recommendation in the context of browsing the web.

6 Conclusions and Future Directions

This paper introduced the notion of context and how it can be utilized within recommender systems. We presented a user model based on research into memory models developed in cognitive science. The user model consists of short and long term memory with context playing the role of retrieval cues for retrieving user preference models from previous interactions of the user with the recommender systems that are contextually related to the active interaction. The retrieved preference models are used to enrich the data within the user's short term memory, pertaining to the active user interaction. We hypothesize that such an enrichment of the short term memory will produce more accurate recommendations. We presented three different type of contextual cues that may be used within the user model and provided evidence, using two collaborative filtering based approaches, that our approach does indeed improve recommendation quality, measured using standard evaluation metrics.

We believe that the ideas presented within this paper provide a new direction in recommender system research and will lead to new techniques and improved recommender performance. However a number of important research questions remain unanswered. Firstly, heuristics need to be developed for identifying contextual interactions. Secondly, as new interactions take place it is likely that more contextual states will be encountered by the recommender system. Research needs to be carried out into how these new states can be discovered and what effect this will have on the recommender system as a whole. Thirdly, we have mainly dealt with memory based preference models. The use of preference models that are not memory based raise challenges in terms of how they can be updated as new interactions take place within existing contexts. Also, how the process of enriching short term memory using multiple such models retrieved from long term memory, needs to be studied.

References

1. Billsus, D., Pazzani, M.J.: User modeling for adaptive news access. User Modelling and User-Adapted Interaction 10, 147–180 (2000)
2. Suchman, L.: Plans and Situated Actions. Cambridge University Press, Cambridge (1987)
3. Chi, E.H.: Transient user profiles. In: Proceedings of the Workshop on User Profiling, pp. 521–523 (2004)
4. Smith, S.M.: Remembering in and out of context. Journal of Experimental Psychology: Human Learning and Memory 5, 460–471 (1979)
5. Bartlett, J.C., Santrock, J.: Affect-depedent episodic memory in young children. Child Development 5, 513–518 (1979)
6. Leech, G.: Semantics: The Study of Meaning, 2nd edn. Penguin (1981)
7. Schilit, B., Theimer, M.: Disseminating active map information to mobile hosts. IEEE Network 8, 22–32 (1994)
8. Dourish, P.: What do we talk about when we talk about context. Personal and Ubiquitous Computing 8(1), 19–30 (2004)
9. Mayrhofer, R., Radi, H., Ferscha, A.: Recognizing and predicting context by learning from user behavior. Radiomatics: Journal of Communication Engineering, special issue on Advances in Mobile Multimedia 1(1), 30–42 (2004)
10. Dourish, P., Edwards, W.K., LaMarca, A., Lamping, J., Petersen, K., Salisbury, M., Terry, D.B., Thornton, J.: Extending document management systems with user-specific active properties. ACM Transactions on Information Systems 18(2), 140–170 (2000)
11. Brumitt, B., Meyers, K.J., Kern, A., Shafer, S.: Easyliving: Technologies for intelligent environments. In: Handheld and Ubiquitous Computing (September 2000)
12. Lieberman, H., Selker, T.: Out of context: Computer systems that adapt to, and learn from, context. IBM Systems Journal 39(3 & 4) (2000)
13. Dey, A.K.: Understanding and using context. Personal and Ubiquitous Computing 5(1), 4–7 (2001)
14. Sieg, A., Mobasher, B., Burke, R.: Inferring user's information context: Integrating user profiles and concept hierarchies. In: Proceedings of the 2004 Meeting of the International Federation of Classification Societies (2004)
15. Kraft, R., Maghoul, F., Chang, C.C.: Y!q: Context search at the point of inspiration. In: Proceedings of the ACM Conference on Information and Knowledge Management, pp. 816–823. ACM Press, New York (2005)
16. Cooley, R., Mobasher, B., Srivastava, J.: Data preparation for mining world wide web browsing patterns. Knowledge and Information Systems 1(1) (1999)

17. Jin, X., Zhou, Y., Mobasher, B.: Task-oriented web user modeling for recommendation. In: Ardissono, L., Brna, P., Mitrović, A. (eds.) UM 2005. LNCS (LNAI), vol. 3538, Springer, Heidelberg (2005)
18. Smyth, P.: Clustering using monte carlo cross-validation. In: Proceedings of the 2nd International Conference on Knowledge Discovery and Data Mining, pp. 126–133 (1996)
19. Adomavicius, G., Tuzhilin, A.: Toward the next generation of recommender systems: a survey of the state-of -the-art and possible extensions. IEEE Transactions on Knowledge and Data Engineering 17, 734–749 (2005)
20. Anand, S.S., Mobasher, B.: Intelligent techniques in web personalization. In: Mobasher, B., Anand, S.S. (eds.) ITWP 2003. LNCS (LNAI), vol. 3169, pp. 1–37. Springer, Heidelberg (2005)
21. Burke, R.: Hybrid recommender systems: Survey and experiments. User Modelling and User Adapted Interaction 12(4), 331–370 (2002)
22. Atkinson, R.C., Shiffrin, R.M.: Human memory: A proposed system and its control processes. Psychology of Learning and Motivation 2, 89–195 (1968)
23. Raaijmakers, J.G.W., Shiffrin, R.M.: Sam: A theory of probabilistic search of associative memory. The Psychology of Learning and Motivation 14, 207–262 (1980)
24. Anand, S.S., Kearney, P., Shapcott, M.: Generating semantically enriched user profiles for web personalization. ACM Transactions on Internet Technologies 7(2) (to appear, 2007)
25. Dai, H., Mobasher, B.: A road map to more effective web personalization: Integrating domain knowledge with web usage mining. In: Proceedings of the International Conference on Internet Computing, pp. 58–64 (2003)
26. Munkres, J.: Algorithms for the assignment and transportation problems. Journal of the Society of Industrial and Applied Mathematics 5(1), 32–38 (1957)
27. Kullback, S., Leibler, R.A.: On information and sufficiency. Annals of Mathematical Statistics 22(1), 79–86 (1951)
28. Adomavicius, G., Sankaranarayanan, R., Sen, S., Tuzhilin, A.: Incorporating contextual information in recommender systems using a multidimensional approach. ACM Transactions on Information Systems 23, 103–145 (2005)
29. Berendt, B., Teltzrow, M.: Addressing users' privacy concerns for improving personalization quality: Towards an integration of user studies and algorithm evaluation. In: Intelligent Techniques in Web Personalisation. LNCS (LNAI), Springer, Heidelberg (2005)
30. Sarwar, B.M., Konstan, J.A., Borchers, A., Herlocker, J., Miller, B., Riedl, J.: Using filtering agents to improve prediction quality in the grouplens research collaborative filtering system. In: Computer Supported Cooperative Work, pp. 345–354 (1998)
31. Hayes, P.C.C.: Context boosting collabirative recommendations. Knowledge Based Systems 17, 131–138 (2004)
32. Forbus, K.D., Gentner, D., Law, K.: Mac/fac: A model of similarity-based retrieval. Cognitive Science 19(2), 141–205 (1994)
33. Fabio Gasparetti, A.M.: User profile generation based on a memory retrieval theory. In: Proc. 1st International Workshop on Web Personalization, Recommender Systems and Intelligent User Interfaces (WPRSIUI'05) (2005)
34. Chi, E.H., Pirolli, P., Pitkow, J.E.: The scent of a site: a system for analyzing and predicting information scent, usage, and usability of a web site. In: Proceedings of the ACM Conference on Human Factors in COmputing Systems, pp. 161–168. ACM Press, New York (2000)

Author Index

Lecture Notes in Artificial Intelligence (LNAI)

Vol. 4529: P. Melin, O. Castillo, L.T. Aguilar, J. Kacprzyk, W. Pedrycz (Eds.), Foundations of Fuzzy Logic and Soft Computing. XIX, 830 pages. 2007.

Vol. 4520: M.V. Butz, O. Sigaud, G. Pezzulo, G. Baldassarre (Eds.), Anticipatory Behavior in Adaptive Learning Systems. X, 379 pages. 2007.

Vol. 4511: C. Conati, K. McCoy, G. Paliouras (Eds.), User Modeling 2007. XVI, 487 pages. 2007.

Vol. 4509: Z. Kobti, D. Wu (Eds.), Advances in Artificial Intelligence. XII, 552 pages. 2007.

Vol. 4496: N.T. Nguyen, A. Grzech, R.J. Howlett, L.C. Jain (Eds.), Agent and Multi-Agent Systems: Technologies and Applications. XXI, 1046 pages. 2007.

Vol. 4483: C. Baral, G. Brewka, J. Schlipf (Eds.), Logic Programming and Nonmonotonic Reasoning. IX, 327 pages. 2007.

Vol. 4482: A. An, J. Stefanowski, S. Ramanna, C.J. Butz, W. Pedrycz, G. Wang (Eds.), Rough Sets, Fuzzy Sets, Data Mining and Granular Computing. XIV, 585 pages. 2007.

Vol. 4481: J. Yao, P. Lingras, W.-Z. Wu, M. Szczuka, N.J. Cercone, D. Ślęzak (Eds.), Rough Sets and Knowledge Technology. XIV, 576 pages. 2007.

Vol. 4476: V. Gorodetsky, C. Zhang, V.A. Skormin, L. Cao (Eds.), Autonomous Intelligent Systems: Multi-Agents and Data Mining. XIII, 323 pages. 2007.

Vol. 4456: Y. Wang, Y.-m. Cheung, H. Liu (Eds.), Computational Intelligence and Security. XXIII, 1118 pages. 2007.

Vol. 4455: S. Muggleton, R. Otero, A. Tamaddoni-Nezhad (Eds.), Inductive Logic Programming. XII, 456 pages. 2007.

Vol. 4452: M. Fasli, O. Shehory (Eds.), Agent-Mediated Electronic Commerce. VIII, 249 pages. 2007.

Vol. 4451: T.S. Huang, A. Nijholt, M. Pantic, A. Pentland (Eds.), Artifical Intelligence for Human Computing. XVI, 359 pages. 2007.

Vol. 4441: C. Müller (Ed.), Speaker Classification. X, 309 pages. 2007.

Vol. 4438: L. Maicher, A. Sigel, L.M. Garshol (Eds.), Leveraging the Semantics of Topic Maps. X, 257 pages. 2007.

Vol. 4434: G. Lakemeyer, E. Sklar, D.G. Sorrenti, T. Takahashi (Eds.), RoboCup 2006: Robot Soccer World Cup X. XIII, 566 pages. 2007.

Vol. 4429: R. Lu, J.H. Siekmann, C. Ullrich (Eds.), Cognitive Systems. X, 161 pages. 2007.

Vol. 4428: S. Edelkamp, A. Lomuscio (Eds.), Model Checking and Artificial Intelligence. IX, 185 pages. 2007.

Vol. 4426: Z.-H. Zhou, H. Li, Q. Yang (Eds.), Advances in Knowledge Discovery and Data Mining. XXV, 1161 pages. 2007.

Vol. 4411: R.H. Bordini, M. Dastani, J. Dix, A.E.F. Seghrouchni (Eds.), Programming Multi-Agent Systems. XIV, 249 pages. 2007.

Vol. 4410: A. Branco (Ed.), Anaphora: Analysis, Algorithms and Applications. X, 191 pages. 2007.

Vol. 4399: T. Kovacs, X. Llorà, K. Takadama, P.L. Lanzi, W. Stolzmann, S.W. Wilson (Eds.), Learning Classifier Systems. XII, 345 pages. 2007.

Vol. 4390: S.O. Kuznetsov, S. Schmidt (Eds.), Formal Concept Analysis. X, 329 pages. 2007.

Vol. 4389: D. Weyns, H. Van Dyke Parunak, F. Michel (Eds.), Environments for Multi-Agent Systems III. X, 273 pages. 2007.

Vol. 4386: P. Noriega, J. Vázquez-Salceda, G. Boella, O. Boissier, V. Dignum, N. Fornara, E. Matson (Eds.), Coordination, Organizations, Institutions, and Norms in Agent Systems II. XI, 373 pages. 2007.

Vol. 4384: T. Washio, K. Satoh, H. Takeda, A. Inokuchi (Eds.), New Frontiers in Artificial Intelligence. IX, 401 pages. 2007.

Vol. 4371: K. Inoue, K. Satoh, F. Toni (Eds.), Computational Logic in Multi-Agent Systems. X, 315 pages. 2007.

Vol. 4369: M. Umeda, A. Wolf, O. Bartenstein, U. Geske, D. Seipel, O. Takata (Eds.), Declarative Programming for Knowledge Management. X, 229 pages. 2006.

Vol. 4343: C. Müller (Ed.), Speaker Classification I. X, 355 pages. 2007.

Vol. 4342: H. de Swart, E. Orłowska, G. Schmidt, M. Roubens (Eds.), Theory and Applications of Relational Structures as Knowledge Instruments II. X, 373 pages. 2006.

Vol. 4335: S.A. Brueckner, S. Hassas, M. Jelasity, D. Yamins (Eds.), Engineering Self-Organising Systems. XII, 212 pages. 2007.

Vol. 4334: B. Beckert, R. Hähnle, P.H. Schmitt (Eds.), Verification of Object-Oriented Software. XXIX, 658 pages. 2007.

Vol. 4333: U. Reimer, D. Karagiannis (Eds.), Practical Aspects of Knowledge Management. XII, 338 pages. 2006.

Vol. 4327: M. Baldoni, U. Endriss (Eds.), Declarative Agent Languages and Technologies IV. VIII, 257 pages. 2006.

Vol. 4314: C. Freksa, M. Kohlhase, K. Schill (Eds.), KI 2006: Advances in Artificial Intelligence. XII, 458 pages. 2007.

Vol. 4304: A. Sattar, B.-h. Kang (Eds.), AI 2006: Advances in Artificial Intelligence. XXVII, 1303 pages. 2006.

Vol. 4303: A. Hoffmann, B.-h. Kang, D. Richards, S. Tsumoto (Eds.), Advances in Knowledge Acquisition and Management. XI, 259 pages. 2006.

Vol. 4293: A. Gelbukh, C.A. Reyes-Garcia (Eds.), MICAI 2006: Advances in Artificial Intelligence. XXVIII, 1232 pages. 2006.

Vol. 4289: M. Ackermann, B. Berendt, M. Grobelnik, A. Hotho, D. Mladenič, G. Semeraro, M. Spiliopoulou, G. Stumme, V. Svátek, M. van Someren (Eds.), Semantics, Web and Mining. X, 197 pages. 2006.

Vol. 4285: Y. Matsumoto, R.W. Sproat, K.-F. Wong, M. Zhang (Eds.), Computer Processing of Oriental Languages. XVII, 544 pages. 2006.